Genetically Modified Organisms

Genetically Modified Organisms

Editor: Leland Carlyle

www.callistoreference.com

Callisto Reference,
118-35 Queens Blvd., Suite 400,
Forest Hills, NY 11375, USA

Visit us on the World Wide Web at:
www.callistoreference.com

ISBN: 978-1-64116-005-6 (Hardback)

Cataloging-in-Publication Data

Genetically modified organisms / edited by Leland Carlyle.
    p. cm.
Includes bibliographical references and index.
ISBN 978-1-64116-005-6
1. Transgenic organisms. 2. Genetic engineering. I. Carlyle, Leland.
QH442.6 .G46 2018
636.082 1--dc23

Table of Contents

Permissions

Index

Preface

The organisms whose genetic traits are changed using the genetic engineering techniques are known as genetically modified organisms. These organisms are better known as 'transgenic organisms' because using other genetic systems from different organisms alters their genetic system. The most commonly modified organisms are viruses, microbes, bacteria, cisgenic plants, transgenic plants, genetically modified crops, etc. This book elucidates the concepts and innovative models around prospective developments with respect to genetically modified organisms. Those in search of information to further their knowledge will be greatly assisted by it. Through this textbook, we attempt to further enlighten the readers about the new concepts in this field.

To facilitate a deeper understanding of the contents of this book a short introduction of every chapter is written below:

Chapter 1- Genetic engineering is the process of modifying genomes. It is applied in various fields such as agriculture, medicine and research. Pharming, transgene, transgenesis, horizontal gene transfer and molecular cloning are some of the topics discussed in this section. This is an introductory chapter which will introduce briefly all the significant aspects of genetic engineering.

Chapter 2- Genetic engineering has several varying techniques which are used in modifying the genomes of an organism. The methods and techniques used in genetic engineering are gene knockin, gene targeting, gene knockout, electroporation and gene knockdown. This chapter discusses the methods and techniques of genetic engineering in a critical manner providing key analysis to the subject matter.

Chapter 3- Genetically modified organisms are organisms whose genome has been altered. The modification consists of insertion, deletion and the mutation of genes. Genetically modified bacteria, genetically modified virus, genetically modified mammal, genetically modified insect and genetically modified fish are the types of organisms that humans have genetically modified. This section is an overview of the subject matter incorporating all the major aspects of genetically modified organisms.

Chapter 4- The crops or plants that are modified by using genetic engineering techniques are known as genetically modified plants. The types of modifications used are transgenic, cisgenic and subgenic. Some of the genetically engineered plants are potato, rice, soybean, tomato, wheat and sugar beet. The topics discussed in the chapter are of great importance to broaden the existing knowledge on genetically modified plants.

Chapter 5- Genetic testing recognizes variations that occur in chromosomes, genes or proteins. These tests are done to study genes to understand the possibility of specific diseases or disorders that can occur in the organism in population ecology and stock quality in agriculture. In order to completely understand genetics it is necessary to understand the processes related to it. The following chapter elucidates the varied processes and mechanisms associated with this area of study.

Chapter 6- Transformation is the alteration of cells resulting from the incorporation of genetic material from its natural surroundings. For the process of transformation to take place, the bacteria must be in a state of competence. The subject deals with themes like natural competence, agrobacterium, viral transformation and transformation efficiency. This chapter will provide an integrated understanding of transformation genetics.

I owe the completion of this book to the never-ending support of my family, who supported me throughout the project.

Editor

An Introduction to Genetic Engineering

Genetic engineering is the process of modifying genomes. It is applied in various fields such as agriculture, medicine and research. Pharming, transgene, transgenesis, horizontal gene transfer and molecular cloning are some of the topics discussed in this section. This is an introductory chapter which will introduce briefly all the significant aspects of genetic engineering.

Genetic Engineering

Genetic engineering, also called genetic modification, is the direct manipulation of an organism's genome using biotechnology. It is a set of technologies used to change the genetic makeup of cells, including the transfer of genes within and across species boundaries to produce improved or novel organisms. New DNA is obtained by either isolating and copying the genetic material of interest using molecular cloning methods or by artificially synthesizing the DNA. A construct is usually created and used to insert this DNA into the host organism. As well as inserting genes, the process can also be used to remove, or "knock out", genes. The new DNA can be inserted randomly, or targeted to a specific part of the genome.

An organism that is generated through genetic engineering is considered to be genetically modified (GM) and the resulting entity is a genetically modified organism (GMO). The first GMOs were bacteria generated in 1973 and the first GM animals were mice in 1974. Insulin-producing bacteria were commercialized in 1982 and genetically modified food has been sold since 1994. GloFish, the first GMO designed as a pet, was sold in the United States in December 2003.

Genetic engineering techniques have been applied in numerous fields including research, agriculture, industrial biotechnology, and medicine. Enzymes used in laundry detergent and medicines such as insulin and human growth hormone are now manufactured in GM cells, experimental GM cell lines and GM animals such as mice or zebrafish are being used for research purposes, and genetically modified crops have been commercialized.

Definition

Genetic engineering alters the genetic make-up of an organism using techniques that remove heritable material or that introduce DNA prepared outside the organism either directly into the host or into a cell that is then fused or hybridized with the host. This involves using recombinant nucleic acid (DNA or RNA) techniques to form new combinations of heritable genetic material followed by the incorporation of that material either indirectly through a vector system or directly through micro-injection, macro-injection and micro-encapsulation techniques. More broadly the definition of genetic engineering has be used to describe selective breeding and other means of artificial selection.

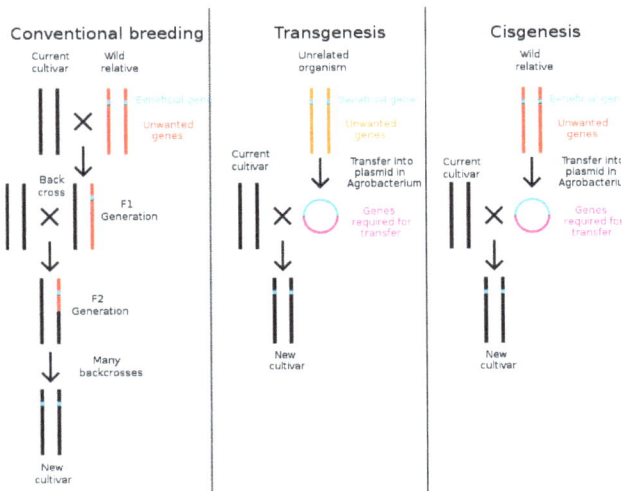

Comparison of conventional plant breeding with transgenic and cisgenic genetic modification.

Genetic engineering does not normally include traditional animal and plant breeding, in vitro fertilisation, induction of polyploidy, mutagenesis and cell fusion techniques that do not use recombinant nucleic acids or a genetically modified organism in the process. Cloning and stem cell research, although not considered genetic engineering, are closely related and genetic engineering can be used within them. Synthetic biology is an emerging discipline that takes genetic engineering a step further by introducing artificially synthesized material from raw materials into an organism.

Plants, animals or micro organisms that have been changed through genetic engineering are termed genetically modified organisms or GMOs. If genetic material from another species is added to the host, the resulting organism is called transgenic. If genetic material from the same species or a species that can naturally breed with the host is used the resulting organism is called cisgenic. Genetic engineering can also be used to remove genetic material from the target organism, creating a gene knockout organism. In Europe genetic modification is synonymous with genetic engineering while within the United States of America and Canada it can also refer to conventional breeding methods.

History

Humans have altered the genomes of species for thousands of years through selective breeding, or artificial selection as contrasted with natural selection, and more recently through mutagenesis. Genetic engineering as the direct manipulation of DNA by humans outside breeding and mutations has only existed since the 1970s. The term "genetic engineering" was first coined by Jack Williamson in his science fiction novel *Dragon's Island*, published in 1951 – one year before DNA's role in heredity was confirmed by Alfred Hershey and Martha Chase, and two years before James Watson and Francis Crick showed that the DNA molecule has a double-helix structure – though the general concept of direct genetic manipulation was explored in rudimentary form in Stanley G. Weinbaum's 1936 science fiction story *Proteus Island*.

In 1972, Paul Berg created the first recombinant DNA molecules by combining DNA from the monkey virus SV40 with that of the lambda virus. In 1973 Herbert Boyer and Stanley Cohen created the first transgenic organism by inserting antibiotic resistance genes into the plasmid of an *E. coli* bacterium. A year later Rudolf Jaenisch created a transgenic mouse by introducing foreign DNA into its embryo, making it the world's first transgenic animal. These achievements led to concerns in the scientific community about potential risks from genetic engineering, which were first discussed in depth at the Asilomar Conference in 1975. One of the main recommendations from this meeting was that government oversight of recombinant DNA research should be established until the technology was deemed safe.

In 1974 Rudolf Jaenisch created the first GM animal

In 1976 Genentech, the first genetic engineering company, was founded by Herbert Boyer and Robert Swanson and a year later the company produced a human protein (somatostatin) in *E.coli*. Genentech announced the production of genetically engineered human insulin in 1978. In 1980, the U.S. Supreme Court in the *Diamond v. Chakrabarty* case ruled that genetically altered life could be patented. The insulin produced by bacteria, branded humulin, was approved for release by the Food and Drug Administration in 1982.

In 1983, a biotech company, Advanced Genetic Sciences (AGS) applied for U.S. government authorization to perform field tests with the ice-minus strain of *P. syringae* to protect crops from frost, but environmental groups and protestors delayed the field tests for four years with legal challenges. In 1987, the ice-minus strain of *P. syringae* became the first genetically modified organism (GMO) to be released into the environment when a strawberry field and a potato field in California were sprayed with it. Both test fields were attacked by activist groups the night before the tests occurred: "The world's first trial site attracted the world's first field trasher".

The first field trials of genetically engineered plants occurred in France and the USA in 1986, tobacco plants were engineered to be resistant to herbicides. The People's Republic of China was the first country to commercialize transgenic plants, introducing a virus-resistant tobacco in 1992. In 1994 Calgene attained approval to commercially release the Flavr Savr tomato, a tomato engineered to have a longer shelf life. In 1994, the European Union approved tobacco engineered to be resistant to the herbicide bromoxynil, making it the first genetically engineered crop commercialized in Europe. In 1995, Bt Potato was approved safe by the Environmental Protection Agency, after having been approved by the FDA, making it the first pesticide producing crop to be approved in the USA. In 2009 11 transgenic crops were grown commercially in 25 countries, the largest of which by area grown were the USA, Brazil, Argentina, India, Canada, China, Paraguay and South Africa.

In 2010, scientists at the J. Craig Venter Institute created the first synthetic genome and inserted it into an empty bacterial cell. The resulting bacterium, named Synthia, could replicate and produce proteins. Four years later this was taken a step further when bacterium was developed that replicated a plasmid containing a unique base pair, creating the first organism engineered to use an expanded genetic alphabet. In 2013, researcher reported the first use of clustered regularly interspaced short palindromic repeats (CRISPR), a technique which can be used to easily alter the genome of almost any organism.

Process

Creating a GMO is a multi-step process. Genetic engineers must first choose what gene they wish to insert into the organism. This is driven by what the aim is for the resultant organism and is built on earlier research. Screens can be carried out to determine potential genes and further tests then used to identify the best candidates. The development of microarrays, transcriptomes and genome sequencing has made it much easier to find suitable genes. Luck also plays its part, the round-up ready gene was discovered after scientist noticed a bacteria thriving in the presence of the herbicide.

Gene Isolation and Cloning

The next step is to isolate the candidate gene. The cell containing the gene is opened and the DNA is purified. The gene is separated by using restriction enzymes to cut the DNA into fragments or Polymerase chain reaction (PCR) to amplify up the gene segment. These segments can then be extracted through gel electrophoresis. If the chosen gene or the donor organism's genome has been well studied it may already be accessible from a genetic library. If the DNA sequence is known, but no copies of the gene are available, it can also be artificially synthesized. Once isolated the gene is ligated into a plasmid that is then inserted into a bacteria. The plasmid is replicated when the bacteria divide, ensuring unlimited copies of the gene are available.

Before the gene is inserted into the target organism it must be combined with other genetic elements. These include a promoter and terminator region, which initiate and end transcription. A selectable marker gene is added, which in most cases confers antibiotic resistance, so researchers can easily determine which cells have been successfully transformed. The gene can also be modified at this stage for better expression or effectiveness. These manipulations are carried out using recombinant DNA techniques, such as restriction digests, ligations and molecular cloning.

Inserting DNA into the Host Genome

There are a number of techniques available for inserting the gene into the host genome. Some bacteria can naturally take up foreign DNA. This ability can be induced in other bacteria via stress (e.g. thermal or electric shock), which increases the cell membrane's permeability to DNA; up-taken DNA can either integrate with the genome or exist as extrachromosomal DNA. DNA is generally inserted into animal cells using microinjection, where it can be injected through the cell's nuclear envelope directly into the nucleus, or through the use of viral vectors.

In plants the DNA is generally inserted using *Agrobacterium*-mediated recombination, taking advantage of the *Agrobacterium*s T-DNA sequence that allows natural insertion of genetic material into plant cells. Another method is biolistics, where particles of gold or tungsten are coated with DNA and then shot into young plant cells or plant embryos. Another transformation method for plant and animal cells is electroporation. This involves subjecting the cell to an electric shock, which can make the cell membrane permeable to plasmid DNA. Due to the damage caused to the cells and DNA the transformation efficiency of biolistics and electroporation is lower than agrobacterial mediated transformation and microinjection.

A. tumefaciens attaching itself to a carrot cell

As only a single cell is transformed with genetic material, the organism must be regenerated from that single cell. As bacteria consist of a single cell and reproduce clonally regeneration is not necessary. In plants this is accomplished through the use of tissue culture. In animals it is necessary to ensure that the inserted DNA is present in the embryonic stem cells. Selectable markers are used to easily differentiate transformed from untransformed cells. These markers are usually present in the transgenic organism, although a number of strategies have been developed that can remove the selectable marker from the mature transgenic plant.

Further testing using PCR, Southern hybridization, and DNA sequencing is conducted to confirm that an organism contains the new gene. These tests can also confirm the chromosomal location and copy number of the inserted gene. The presence of the gene does not guarantee it will be expressed at appropriate levels in the target tissue so methods that look for and measure the gene products (RNA and protein) are also used. These include northern hybridization, quantitative RT-PCR, Western blot, immunofluorescence, ELISA and phenotypic analysis. All offspring from the first generation will be heterozygous for the inserted gene and must be mated together to produce a homozygous animal. For stable transformation the gene should be passed to the offspring in a Mendelian inheritance pattern, so the organism's offspring are also studied.

The new genetic material can be inserted randomly within the host genome or targeted to a specific location. The technique of gene targeting uses homologous recombination to make desired changes to a specific endogenous gene. This tends to occur at a relatively low frequency in plants and animals and generally requires the use of selectable markers. The frequency of gene targeting can be greatly enhanced through genome editing. Genome editing uses artificially engineered nucleases that create specific double-stranded breaks at desired locations in the genome, and use the cell's endogenous mechanisms to repair the induced break by the natural processes of homologous recombination and nonhomologous end-joining. There are currently four families of engineered nucleases: meganucleases, zinc finger nucleases, transcription activator-like effector nucleases (TALENs), and the Cas9-guideRNA system (adapted from CRISPR). In addition to enhancing gene targeting, engineered nucleases can also be used to introduce mutations at endogenous genes that generate a gene knockout.

Applications

Genetic engineering has applications in medicine, research, industry and agriculture and can be used on a wide range of plants, animals and micro organisms. Bacteria, the first organisms to be genetically modified, can have plasmid DNA inserted containing new genes that code for medicines or enzymes that process food and other substrates. Plants have been modified for insect protection, herbicide resistance, virus resistance, enhanced nutrition, tolerance to environmental pressures and the production of edible vaccines. Most commercialised GMOs are insect resistant and/or herbicide tolerant crop plants. Genetically modified animals have been used for research, model animals and the production of agricultural or pharmaceutical products. The genetically modified animals include animals with genes knocked out, increased susceptibility to disease, hormones for extra growth and the ability to express proteins in their milk.

Medicine

Genetic engineering has many applications to medicine that include the manufacturing of drugs, creation of model animals that mimic human conditions and gene therapy. One of the earliest uses of genetic engineering was to mass-produce human insulin in bacteria. This application has now been applied to, human growth hormones, follistim (for treating infertility), human albumin, monoclonal antibodies, antihemophilic factors, vaccines and many other drugs. Mouse hybridomas, cells fused together to create monoclonal antibodies, have been humanised through genetic engineering to create human monoclonal antibodies. Genetically engineered viruses are being developed that can still confer immunity, but lack the infectious sequences.

Genetic engineering is used to create animal models of human diseases. Genetically modified mice are the most common genetically engineered animal model. They have been used to study and model cancer (the oncomouse), obesity, heart disease, diabetes, arthritis, substance abuse, anxiety, aging and Parkinson disease. Potential cures can be tested against these mouse models. Also genetically modified pigs have been bred with the aim of increasing the success of pig to human organ transplantation.

Gene therapy is the genetic engineering of humans, generally by replacing defective genes with effective ones. Clinical research using somatic gene therapy has been conducted with several diseases, including X-linked SCID, chronic lymphocytic leukemia (CLL), and Parkinson's disease. In 2012, Glybera became the first gene therapy treatment to be approved for clinical use. germline gene therapy would result in any change being inheritable, which has raised concerns within the scientific community. In 2015, CRISPR was used to edit the DNA of non-viable human embryos, leading scientists of major world academies to called for a moratorium on inheritable human genome edits.

There are also concerns that the technology could be used not just for treatment, but for enhancement, modification or alteration of a human beings' appearance, adaptability, intelligence, character or behavior. The distinction between cure and enhancement can also be difficult to establish.

Research

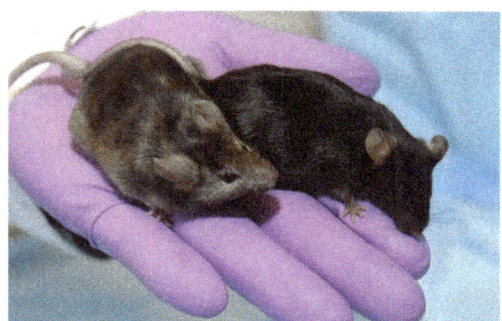

Knockout mice

Genetic engineering is an important tool for natural scientists. Genes and other genetic information from a wide range of organisms are transformed into bacteria for storage and modification, creating genetically modified bacteria in the process. Bacteria are cheap, easy to grow, clonal, multiply quickly, relatively easy to transform and can be stored at -80 °C almost indefinitely. Once a gene is isolated it can be stored inside the bacteria providing an unlimited supply for research.

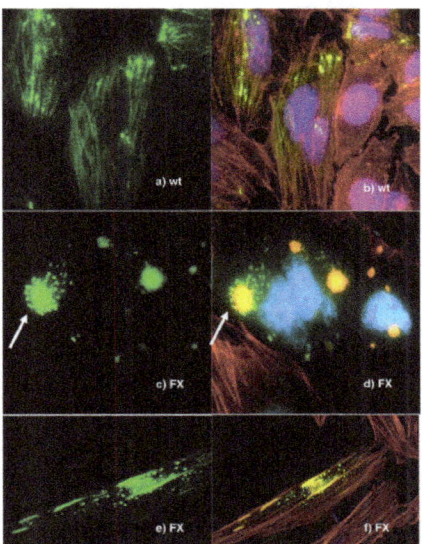

Human cells in which some proteins are fused with green fluorescent protein to allow them to be visualised

Organisms are genetically engineered to discover the functions of certain genes. This could be the effect on the phenotype of the organism, where the gene is expressed or what other genes it interacts with. These experiments generally involve loss of function, gain of function, tracking and expression.

- Loss of function experiments, such as in a gene knockout experiment, in which an organism is engineered to lack the activity of one or more genes. A knockout experiment involves the creation and manipulation of a DNA construct *in vitro*, which, in a simple knockout, consists of a copy of the desired gene, which has been altered such that it is non-functional. Embryonic stem cells incorporate the altered gene, which replaces the already present functional copy. These stem cells are injected into blastocysts, which are implanted into surrogate mothers. This allows the experimenter to analyze the defects caused by this mutation and thereby determine the role of particular genes. It is used especially frequently in developmental biology. Another method, useful in organisms such as Drosophila (fruit fly), is to induce mutations in a large population and then screen the progeny for the desired mutation. A similar process can be used in both plants and prokaryotes. Loss of function tells whether or not a protein is required for a function, but does not always mean it's sufficient, especially if a function requires multiple proteins and is lost if one protein is missing.

- Gain of function experiments, the logical counterpart of knockouts. These are sometimes performed in conjunction with knockout experiments to more finely establish the function of the desired gene. The process is much the same as that in knockout engineering, except that the construct is designed to increase the function of the gene, usually by providing extra copies of the gene or inducing synthesis of the protein more frequently. Gain of function is used to tell whether or not a protein is sufficient for a function, but does not always mean it's required, especially when dealing with genetic or functional redundancy.

- Tracking experiments, which seek to gain information about the localization and interaction of the desired protein. One way to do this is to replace the wild-type gene with a 'fusion' gene, which is a juxtaposition of the wild-type gene with a reporting element such as green fluorescent protein (GFP) that will allow easy visualization of the products of the genetic modification. While this is a useful technique, the manipulation can destroy the function of the gene, creating secondary effects and possibly calling into question the results of the experiment. More sophisticated techniques are now in development that can track protein products without mitigating their function, such as the addition of small sequences that will serve as binding motifs to monoclonal antibodies.

- Expression studies aim to discover where and when specific proteins are produced. In these experiments, the DNA sequence before the DNA that codes for a protein, known as a gene's promoter, is reintroduced into an organism with the protein coding region replaced by a reporter gene such as GFP or an enzyme that catalyzes the production of a dye. Thus the time and place where a particular protein is produced can be observed. Expression studies can be taken a step further by altering the promoter to find which pieces are crucial for the proper expression of the gene and are actually bound by transcription factor proteins; this process is known as promoter bashing.

Industrial

Organisms can have their cells transformed with a gene coding for a useful protein, such as an enzyme, so that the they will overexpress the desired protein. Mass quantities of the protein can then be manufactured by growing the transformed organism in bioreactor equipment using industrial fermentation, and then purifying the protein. Some genes do not work well in bacteria, so yeast, insect cells or mammalians cells can also be used. These techniques are used to produce medicines such as insulin, human growth hormone, and vaccines, supplements such as tryptophan, aid in the production of food (chymosin in cheese making) and fuels. Other applications with genetically engineered bacteria could involve making them perform tasks outside their natural cycle, such as making biofuels, cleaning up oil spills, carbon and other toxic waste and detecting arsenic in drinking water. Certain genetically modified microbes can also be used in biomining and bioremediation, due to their ability to extract heavy metals from their environment and incorporate them into compounds that are more easily recoverable.

In materials science, a genetically modified virus has been used in an academic lab as a scaffold for assembling a more environmentally friendly lithium-ion battery. Bacteria have also been engineered to function as sensors by expressing a fluorescent protein under certain environmental conditions.

Agriculture

Bt-toxins present in peanut leaves (bottom image) protect it from extensive damage caused by European corn borer larvae (top image).

One of the best-known and controversial applications of genetic engineering is the creation and use of genetically modified crops or genetically modified livestock to produce genetically modified food. Crops have been developed to increase production, increase tolerance to abiotic stresses, alter the composition of the food, or to produce novel products.

The first crops to be realised commercially on a large scale provided protection from insect pests or tolerance to herbicides. Fungal and virus resistant crops have also being developed or are in development. This make the insect and weed management of crops easier and can indirectly increase crop yield. GM crops that directly improve yield by accelerating growth or making the plant more hardy (by improving salt, cold or drought tolerance) are also under development. Salmon have been genetically modified with growth hormones to increase their size.

GMOs have been developed that modify the quality of produce by increasing the nutritional value or providing more industrially useful qualities or quantities. The Amflora potato produces a more industrially useful blend of starches. Cows have been engineered to produce more protein in their milk to facilitate cheese production. Soybeans and canola have been genetically modified to produce more healthy oils. The first commercialised GM food was a tomato that had delayed ripening, increasing its shelf life.

Plants and animals have been engineered to produce materials they do not normally make. Pharming uses crops as bioreactors to produce vaccines, drug intermediates, or the drugs themselves; the useful product is purified from the harvest and then used in the standard pharmaceutical production process. Cows and goats have been engineered to express drugs and other proteins in their milk, and in 2009 the FDA approved a drug produced in goat milk.

Other Applications

Genetic engineering has potential applications in conservation and natural areas management. Gene transfer through viral vectors has been proposed as a means of controlling invasive species as well as vaccinating threatened fauna from disease. Transgenic trees have been suggested as a way to confer resistance to pathogens in wild populations. With the increasing risks of maladaptation in organisms as a result of climate change and other perturbations, facilitated adaptation through gene tweaking could be one solution to reducing extinction risks. Applications of genetic engineering in conservation are thus far mostly theoretical and have yet to be put into practice.

Genetic engineering is also being used to create BioArt. Some bacteria have been genetically engineered to create black and white photographs. Novelty items such as lavender-colored carnations, blue roses, and glowing fish have also been produced through genetic engineering.

Regulation

The regulation of genetic engineering concerns the approaches taken by governments to assess and manage the risks associated with the development and release of GMOs. The development of a regulatory framework began in 1975, at Asilomar, California. The Asilomar meeting recommended a set of voluntary guidelines regarding the use of recombinant technology. As the technology improved USA established a committee at the Office of Science and Technology, which assigned regulatory approval of GM plants to the USDA, FDA and EPA. The Cartagena Protocol on Biosafety, an international treaty that governs the transfer, handling, and use of GMOs, was adopted on 29 January 2000. One hundred and fifty-seven countries are members of the Protocol and many use it as a reference point for their own regulations.

The legal and regulatory status of GM foods varies by country, with some nations banning or restricting them, and others permitting them with widely differing degrees of regulation. Some countries allow the import of GM food with authorization, but either do not allow its cultivation (Russia, Norway, Israel) or have provisions for cultivation, but no GM products are yet produced (Japan, South Korea). Most countries that do not allow for GMO cultivation do permit research. Some of the most marked differences occurring between the USA and Europe. The US policy focuses on the product (not the process), only looks at verifiable scientific risks and uses the concept of substantial equivalence. The European Union by contrast has possibly the most stringent GMO regulations in the world. All GMOs, along with irradiated food, are considered "new food" and subject to extensive, case-by-case, science-based food evaluation by the European Food Safety Authority. The criteria for authorization fall in four broad categories: "safety," "freedom of choice," "labelling," and "traceability." The level of regulation in other countries that cultivate GMOs lie in between Europe and the United States.

The regulation agencies by geographical regions		
Region	**Regulator/s**	**Notes**
USA	USDA, FDA and EPA	
Europe	European Food Safety Authority	
Canada	Health Canada and the Canadian Food Inspection Agency	Based on whether a product has novel features regardless of method of origin
Africa	Common Market for Eastern and Southern Africa	Final decision lies with each individual country.
China	Office of Agricultural Genetic Engineering Biosafety Administration	
India	Institutional Biosafety Committee, Review Committee on Genetic Manipulation and Genetic Engineering Approval Committee	
Argentina	National Agricultural Biotechnology Advisory Committee (environmental impact), the National Service of Health and Agrifood Quality (food safety) and the National Agri-business Direction (effect on trade)	Fnal decision made by the Secretariat of Agriculture, Livestock, Fishery and Food.
Brazil	National Biosafety Technical Commission (environmental and food safety) and the Council of Ministers (commercial and economical issues)	
Australia	Office of the Gene Technology Regulator (overseas all), Therapeutic Goods Administration (GM medicines) and Food Standards Australia New Zealand (GM food).	The individual state governments can then assess the impact of release on markets and trade and apply further legislation to control approved genetically modified products.

One of the key issues concerning regulators is whether GM products should be labeled. The European Commission says that mandatory labeling and traceability are needed to allow for informed choice, avoid potential False advertising and facilitate the withdrawal of products if adverse effects on health or the environment are discovered. The American Medical Association and the American Association for the Advancement of Science say that absent scientific evidence of harm even voluntary labeling is misleading and will falsely alarm consumers". Labeling of GMO products in the marketplace is required in 64 countries. Labeling can be mandatory up to a threshold GM content level (which varies between countries) or voluntary. In Canada and the USA labeling of GM food is voluntary, while in Europe all food (including processed food) or feed which contains greater than 0.9% of approved GMOs must be labelled.

Controversy

Critics have objected to the use of genetic engineering on several grounds, that include ethical, ecological and economic concerns. Many of these concerns involve GM crops and whether food produced from them is safe, whether it should be labeled and what impact growing them will have on the environment. These controversies have led to litigation, international trade disputes, and protests, and to restrictive regulation of commercial products in some countries.

Accusations that scientists are "playing God" and other religious issues have been ascribed to the technology from the beginning. Other ethical issues raised include the patenting of life, the use of intellectual property rights, the level of labeling on products, control of the food supply and the objectivity of the regulatory process. Although doubts have been raised, economically most studies have found growing GM crops to be beneficial to farmers.

Gene flow between GM crops and compatible plants, along with increased use of selective herbicides, can increase the risk of "superweeds" developing. Other environmental concerns involve potential impacts on non-target organisms, including soil microbes, and an increase in secondary and resistant insect pests. Many of the environmental impacts regarding GM crops may take many years to be understood are also evident in conventional agriculture practices. With the commercialisation of genetically modified fish there are concerns over what the environmental consequences will be if they escape.

There are three main concerns over the safety of genetically modified food; whether they may provoke an allergic reaction, whether the genes could transfer from the food into human cells, and whether the genes not approved for human consumption could outcross to other crops. There is a scientific consensus that currently available food derived from GM crops poses no greater risk to human health than conventional food, but that each GM food needs to be tested on a case-by-case basis before introduction. Nonetheless, members of the public are much less likely than scientists to perceive GM foods as safe.

History of Genetic Engineering

Herbert Boyer (pictured) and Stanley Cohen created the first genetically modified organism in 1972

Genetic modification caused by human activity has been occurring since around 12,000 BC, when humans first began to domesticate organisms. Genetic engineering as the direct transfer of DNA from one organism to another was first accomplished by Herbert Boyer and Stanley Cohen in 1972. The first genetically modified animal was a mouse created in 1974 by Rudolf Jaenisch. In 1983 an antibiotic resistant gene was inserted into tobacco, leading to the first genetically engineered plant. Advances followed that allowed scientists to manipulate and add genes to a variety of different organism and induce a range of different effects.

In 1976 the technology was commercialized, with the advent of genetically modified bacteria that produced somatostatin, followed by insulin in 1978. Plants were first commercialized with virus resistant tobacco released in China in 1992. The first genetically modified food was the Flavr Savr tomato marketed in 1994. By 2010, 29 countries had planted commercialized biotech crops. In 2000 a paper published in *Science* introduced golden rice, the first food developed with increased nutrient value.

Agriculture

Genetic engineering is the direct manipulation of an organism's genome using certain biotechnology techniques that have only existed since the 1970s. Human directed genetic manipulation was occurring much earlier, beginning with the domestication of plants and animals through artificial selection. The dog is believed to be the first animal domesticated, possibly arising from a common ancestor of the grey wolf, with archeological evidence dating to about 12,000 BC. Other carnivores domesticated in prehistoric times include the cat, which cohabited with human 9,500 years ago. Archeological evidence suggests sheep, cattle, pigs and goats were domesticated between 9 000 BC and 8 000 BC in the Fertile Crescent.

DNA studies suggested that the dog most likely arose from a common ancestor with the grey wolf

The first evidence of plant domestication comes from emmer and einkorn wheat found in pre-Pottery Neolithic A villages in Southwest Asia dated about 10,500 to 10,100 BC. The Fertile Crescent of Western Asia, Egypt, and India were sites of the earliest planned sowing and harvesting of plants that had previously been gathered in the wild. Independent development of agriculture occurred in northern and southern China, Africa's Sahel, New Guinea and several regions of the Americas. The eight Neolithic founder crops (emmer wheat, einkorn wheat, barley, peas, lentils, bitter vetch, chick peas and flax) had all appeared by about 7000 BC. Horticulture first appears in the Levant during the Chalcolithic period about 6 800 to 6,300 BC. Due to the soft tissues, archeological evidence for early vegetables is scarce. The earliest vegetable remains have been found in Egyptian caves that date back to the 2nd millennium BC.

Selective breeding of domesticated plants was once the main way early farmers shaped organisms to suit their needs. Charles Darwin described three types of selection: methodical selection, wherein humans deliberately select for particular characteristics; unconscious selection, wherein a characteristic is selected simply because it is desirable; and natural selection, wherein a trait that helps an organism survive better is passed on. Early breeding relied on unconscious and natural selection. The introduction of methodical selection is unknown. Common characteristics that were bred into domesticated plants include grains that did not shatter to allow easier harvesting, uniform ripening, shorter lifespans that translate to faster growing, loss of toxic compounds, and productivity. Some plants, like the Banana, were able to be propagated by vegetative cloning. Off-spring often did not contain seeds, and therefore sterile. However, these offspring were usually juicier and larger. Propagation through cloning allows these mutant varieties to be cultivated despite their lack of seeds.

Hybridization was another way that rapid changes in plant's makeup were introduced. It often increased vigor in plants, and combined desirable traits together. Hybridization most likely first occurred when humans first grew similar, yet slightly different plants in close proximity. Triticum aestivum, wheat used in baking bread, is an allopolyploid. Its creation is the result of two separate hybridization events.

Grafting can transfer chloroplasts (specialised DNA in plants that can conduct photosynthesis), mitichondrial DNA and the entire cell nucleus containing the genome to potentially make a new species making grafting a form of natural genetic engineering.

X-rays were first used to deliberately mutate plants in 1927. Between 1927 and 2007, more than 2,540 genetically mutated plant varieties had been produced using x-rays.

Genetics

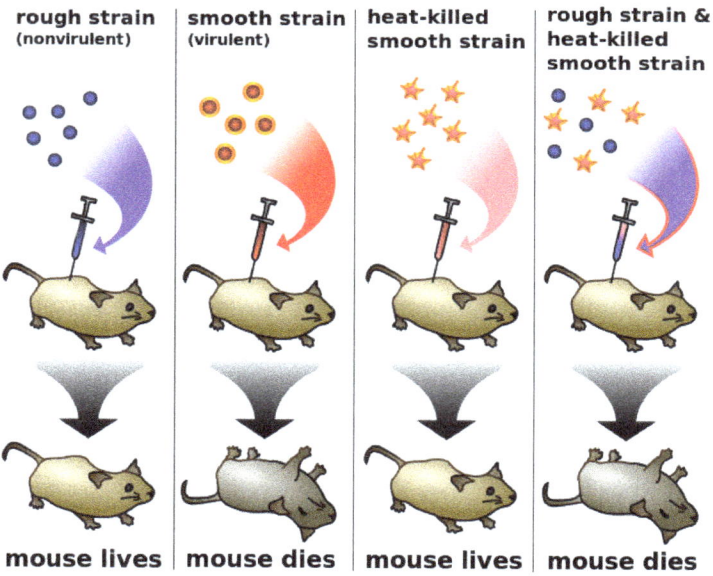

Griffith proved the existence of a "transforming principle", which Avery, MacLeod and
McCarty later showed to be DNA

Various genetic discoveries have been essential in the development of genetic engineering. Genetic inheritance was first discovered by Gregor Mendel in 1865 following experiments crossing peas. Although largely ignored for 34 years he provided the first evidence of hereditary segregation and independent assortment. In 1889 Hugo de Vries came up with the name "(pan)gene" after postulating that particles are responsible for inheritance of characteristics and the term "genetics" was coined by William Bateson in 1905. In 1928 Frederick Griffith proved the existence of a "transforming principle" involved in inheritance, which Avery, MacLeod and McCarty later (1944) identified as DNA. Edward Lawrie Tatum and George Wells Beadle developed the central dogma that genes code for proteins in 1941. The double helix structure of DNA was identified by James Watson and Francis Crick in 1953.

As well as discovering how DNA works, tools had to be developed that allowed it to be manipulated. In 1970 Hamilton Smiths lab discovered restriction enzymes that allowed DNA to be cut at specific places and separated out on an electrophoresis gel. This enabled scientists to isolate genes from an organism's genome. DNA ligases, that join broken DNA together, had been discovered earlier in 1967 and by combining the two enzymes it was possible to "cut and paste" DNA sequences to create recombinant DNA. Plasmids, discovered in 1952, became important tools for transferring information between cells and replicating DNA sequences. Frederick Sanger developed a method for sequencing DNA in 1977, greatly increasing the genetic information available to researchers. Polymerase chain reaction (PCR), developed by Kary Mullis in 1983, allowed small sections of DNA to be amplified and aided identification and isolation of genetic material.

As well as manipulating the DNA, techniques had to be developed for its insertion (known as transformation) into an organism's genome. Griffiths experiment had already shown that some bacteria had the ability to naturally take up and express foreign DNA. Artificial competence was induced in *Escherichia coli* in 1970 when Morton Mandel and Akiko Higa showed that it could take up bacteriophage λ after treatment with calcium chloride solution ($CaCl_2$). Two years later, Stanley Cohen showed that $CaCl_2$ treatment was also effective for uptake of plasmid DNA. Transformation using electroporation was developed in the late 1980s, increasing the efficiency and bacterial range. In 1907 a bacterium that caused plant tumors, *Agrobacterium tumefaciens*, was discovered and in the early 1970s the tumor inducing agent was found to be a DNA plasmid called the Ti plasmid. By removing the genes in the plasmid that caused the tumor and adding in novel genes researchers were able to infect plants with *A. tumefaciens* and let the bacteria insert their chosen DNA into the genomes of the plants.

Early Genetically Modified Organisms

Paul Berg created the first recombinant DNA molecules in 1972

In 1972 Paul Berg utilised restriction enzymes and DNA ligases to create the first recombinant DNA molecules. He combined DNA from the monkey virus SV40 with that of the lambda virus. Herbert Boyer and Stanley Norman Cohen took Berg's work a step further and introduced recombinant DNA into a bacterial cell. Cohen was researching plasmids, while Boyers work involved restriction enzymes. They recognised the complementary nature of their work and teamed up in 1972. Together they found a restriction enzyme that cut the pSC101 plasmid at a single point and were able to insert and ligate a gene that conferred resistance to the kanamycin antibiotic into the gap. Cohen had previously devised a method where bacteria could be induced to take up a plasmid and using this they were able to create a bacteria that survived in the presence of the kanamycin. This represented the first genetically modified organism. They repeated experiments showing that other genes could be expressed in bacteria, including one from the toad Xenopus laevis, the first cross kingdom transformation.

In 1974 Rudolf Jaenisch created a transgenic mouse by introducing foreign DNA into its embryo, making it the world's first transgenic animal. Jaenisch was studying mammalian cells infected with simian virus 40 (SV40) when he happened to read a paper from Beatrice Mintz describing the generation of chimera mice. He took his SV40 samples to Mintz's lab and injected them into early mouse embryos expecting tumours to develop. The mice appeared normal, but after using radioactive probes he discovered that the virus had integrated itself into the mice genome. However the mice did not pass the transgene to their offspring. In 1981 the laboratories of Frank Ruddle, Frank Constantini and Elizabeth Lacy injected purified DNA into a single-cell mouse embryo and showed transmission of the genetic material to subsequent generations.

The first genetically engineered plant was tobacco, reported in 1983. It was developed by Michael W. Bevan, Richard B. Flavell and Mary-Dell Chilton by creating a chimeric gene that joined an antibiotic resistant gene to the T1 plasmid from *Agrobacterium*. The tobacco was infected with *Agrobacterium* transformed with this plasmid resulting in the chimeric gene being inserted into the plant. Through tissue culture techniques a single tobacco cell was selected that contained the gene and a new plant grown from it.

Regulation

The development of genetic engineering technology led to concerns in the scientific community about potential risks. The development of a regulatory framework concerning genetic engineering began in 1975, at Asilomar, California. The Asilomar meeting recommended a set of guidelines regarding the cautious use of recombinant technology and any products resulting from that technology. The Asilomar recommendations were voluntary, but in 1976 the US National Institute of Health (NIH) formed a recombinant DNA advisory committee. This was followed by other regulatory offices (the United States Department of Agriculture (USDA), Environmental Protection Agency (EPA) and Food and Drug Administration (FDA), effectively making all recombinant DNA research tightly regulated in the USA.

In 1982 the Organization for Economic Co-operation and Development (OECD) released a report into the potential hazards of releasing genetically modified organisms into the environment as the first transgenic plants were being developed. As the technology improved and genetically organisms moved from model organisms to potential commercial products the USA established a committee at the Office of Science and Technology (OSTP) to develop mechanisms to regulate the developing technology. In 1986 the OSTP assigned regulatory approval of genetically modified plants in the US to the USDA, FDA and EPA. In the late 1980s and early 1990s, guidance on assessing the safety of genetically engineered plants and food emerged from organizations including the FAO and WHO.

The European Union first introduced laws requiring GMO's to be labelled in 1997. In 2013 Connecticut became the first state to enact a labeling law in the USA, although it would not take effect until other states followed suit.

Research and Medicine

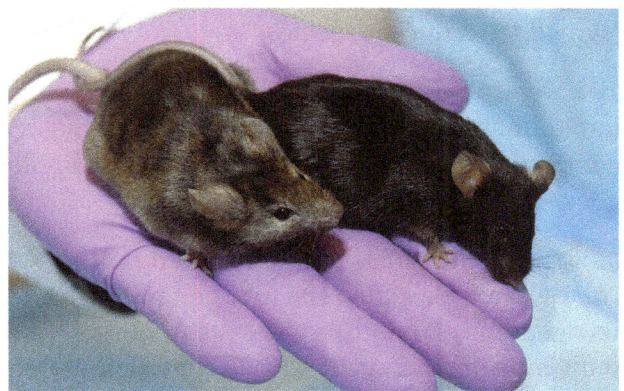

A laboratory mouse in which a gene affecting hair growth has been knocked out (left), is shown next to a normal lab mouse.

The ability to insert, alter or remove genes in model organisms allowed scientists to study the genetic elements of human diseases. Genetically modified mice were created in 1984 that carried cloned oncogenes that predisposed them to developing cancer. The technology has also been used to generate mice with genes knocked out. The first recorded knockout mouse was created by Mario R. Capecchi, Martin Evans and Oliver Smithies in 1989. In 1992 oncomice with tumor suppressor genes knocked out were generated. Creating Knockout rats is much harder and only became possible in 2003.

After the discovery of microRNA in 1993, RNA interference (RNAi) has been used to silence an organism's genes. By modifying an organism to express microRNA targeted to its endogenous genes, researchers have been able to knockout or partially reduce gene function in a range of species. The ability to partially reduce gene function has allowed the study of genes that are lethal when completely knocked out. Other advantages of using RNAi include the availability of inducible and tissue specific knockout. In 2007 microRNA targeted to insect and nematode genes was expressed in plants, leading to suppression when they fed on the transgenic plant, potentially creating a new way to control pests. Targeting endogenous microRNA expression has allowed further fine tuning of gene expression, supplementing the more traditional gene knock out approach.

Genetic engineering has been used to produce proteins derived from humans and other sources in organisms that normally cannot synthesize these proteins. Human insulin-synthesising bacteria were developed in 1979 and were first used as a treatment in 1982. In 1988 the first human antibodies were produced in plants. In 2000 Vitamin A-enriched golden rice, was the first food with increased nutrient value.

Further Advances

As not all plant cells were susceptible to infection by *A. tumefaciens* other methods were developed, including electroporation, micro-injection and particle bombardment with a gene gun (invented in 1987). In the 1980s techniques were developed to introduce isolated chloroplasts back into a plant cell that had its cell wall removed. With the introduction of the gene gun in 1987 it became possible to integrate foreign genes into a chloroplast.

Genetic transformation has become very efficient in some model organism. In 2008 genetically modified seeds were produced in *Arabidopsis thaliana* by simply dipping the flowers in an *Agrobacterium* solution. The range of plants that can be transformed has increased as tissue culture techniques have been developed for different species.

The first transgenic livestock were produced in 1985, by micro-injecting foreign DNA into rabbit, sheep and pig eggs. The first animal to synthesise transgenic proteins in their milk were mice, engineered to produce human tissue plasminogen activator. This technology was applied to sheep, pigs, cows and other livestock.

In 2010 scientists at the J. Craig Venter Institute announced that they had created the first synthetic bacterial genome. The researchers added the new genome to bacterial cells and selected for cells that contained the new genome. To do this the cells undergoes a process called resolution, where during bacterial cell division one new cell receives the original DNA genome of the bacteria, whilst the other receives the new synthetic genome. When this cell replicates it uses the synthetic genome as its template. The resulting bacterium the researchers developed, named Synthia, was the world's first synthetic life form.

In 2014 a bacteria was developed that replicated a plasmid containing an unnatural base pair. This required altering the bacterium so it could import the unnatural nucleotides and then efficiently replicate them. The plasmid retained the unnatural base pairs when it doubled an estimated 99.4% of the time. This is the first organism engineered to use an expanded genetic alphabet.

In 2015 CRISPR and TALENs was used to modify plant genomes. Chinese labs used it to create a fungus-resistant wheat and boost rice yields, while a U.K. group used it to tweak a barley gene that could help produce drought-resistant varieties. When used to precisely remove material from DNA without adding genes from other species, the result is not subject the lengthy and expensive regulatory process associated with GMOs. While CRISPR may use foreign DNA to aid the editing process, the second generation of edited plants contain none of that DNA. Researchers celebrated the acceleration because it may allow them to "keep up" with rapidly evolving pathogens. The U.S. Department of Agriculture stated that some examples of gene-edited corn, potatoes and soybeans are not subject to existing regulations. As of 2016 other review bodies had yet to make statements.

Commercialisation

In 1976 Genentech, the first genetic engineering company was founded by Herbert Boyer and Robert Swanson and a year later and the company produced a human protein (somatostatin) in *E.coli*. Genentech announced the production of genetically engineered human insulin in 1978. In 1980 the U.S. Supreme Court in the Diamond v. Chakrabarty case ruled that genetically altered life could be patented. The insulin produced by bacteria, branded humulin, was approved for release by the Food and Drug Administration in 1982.

In 1983 a biotech company, Advanced Genetic Sciences (AGS) applied for U.S. government authorization to perform field tests with the ice-minus strain of *P. syringae* to protect crops from frost, but environmental groups and protestors delayed the field tests for four years with legal challenges. In 1987 the ice-minus strain of *P. syringae* became the first genetically modified organism (GMO) to be released into the environment when a strawberry field and a potato field in California were sprayed with it. Both test fields were attacked by activist groups the night before the tests occurred: "The world's first trial site attracted the world's first field trasher".

The first genetically modified crop plant was produced in 1982, an antibiotic-resistant tobacco plant. The first field trials of genetically engineered plants occurred in France and the USA in 1986, tobacco plants were engineered to be resistant to herbicides. In 1987 Plant Genetic Systems, founded by Marc Van Montagu and Jeff Schell, was the first company to genetically engineer insect-resistant plants by incorporating genes that produced insecticidal proteins from Bacillus thuringiensis (Bt) into tobacco.

Genetically modified microbial enzymes were the first application of genetically modified organisms in food production and were approved in 1988 by the US Food and Drug Administration. In the early 1990s, recombinant chymosin was approved for use in several countries. Cheese had typically been made using the enzyme complex rennet that had been extracted from cows' stomach lining. Scientists modified bacteria to produce chymosin, which was also able to clot milk, resulting in cheese curds. The People's Republic of China was the first country to commercialize transgenic plants, introducing a virus-resistant tobacco in 1992. In 1994 Calgene attained approval to commercially release the Flavr Savr tomato, a tomato engineered to have a longer shelf life. Also in 1994, the European Union approved tobacco engineered to be resistant to the herbicide bromoxynil, making it the first genetically engineered crop commercialized in Europe. In 1995 Bt Potato was approved safe by the Environmental Protection Agency, after having been approved by the FDA, making it the first pesticide producing crop to be approved in the USA. In 1996 a total of 35 approvals had been granted to commercially grow 8 transgenic crops and one flower crop (carnation), with 8 different traits in 6 countries plus the EU.

By 2010, 29 countries had planted commercialized biotech crops and a further 31 countries had granted regulatory approval for transgenic crops to be imported. In 2013 Robert Fraley (Monsanto's executive vice president and chief technology officer), Marc Van Montagu and Mary-Dell Chilton were awarded the World Food Prize for improving the "quality, quantity or availability" of food in the world.

The first genetically modified animal to be commercialised was the GloFish, a Zebra fish with a fluorescent gene added that allows it to glow in the dark under ultraviolet light. The first genetically modified animal to be approved for food use was AquAdvantage salmon in 2015. The salmon were transformed with a growth hormone-regulating gene from a Pacific Chinook salmon and a promoter from an ocean pout enabling it to grow year-round instead of only during spring and summer.

Opposition

Opposition and support for the use of genetic engineering has existed since the technology was developed. After Arpad Pusztai went public with research he was conducting in 1998 the public opposition to genetically modified food increased. Opposition continued following controversial and publicly debated papers published in 1999 and 2013 that claimed negative environmental and health impacts from genetically modified crops.

Pharming (Genetics)

Pharming, a portmanteau of "farming" and "pharmaceutical", refers to the use of genetic engineering to insert genes that code for useful pharmaceuticals into host animals or plants that would otherwise not express those genes, thus creating a genetically modified organism (GMO). Pharming is also known as molecular farming, molecular pharming or biopharming.

The products of pharming are recombinant proteins or their metabolic products. Recombinant proteins are most commonly produced using bacteria or yeast in a bioreactor, but pharming offers the advantage to the producer that it does not require expensive infrastructure, and production capacity can be quickly scaled to meet demand, at greatly reduced cost.

History

The first recombinant plant-derived protein (PDP) was human serum albumin, initially produced in 1990 in transgenic tobacco and potato plants. Open field growing trials of these crops began in the United States in 1992 and have taken place every year since. While the United States Department of Agriculture has approved planting of pharma crops in every state, most testing has taken place in Hawaii, Nebraska, Iowa, and Wisconsin.

In the early 2000s, the pharming industry was robust. Proof of concept has been established for the production of many therapeutic proteins, including antibodies, blood products, cytokines, growth factors, hormones, recombinant enzymes and human and veterinary vaccines. By 2003 several PDP products for the treatment of human diseases were under development by nearly 200 biotech companies, including recombinant gastric lipase for the treatment of cystic fibrosis, and antibodies for the prevention of dental caries and the treatment of non-Hodgkin's lymphoma.

Several proteins were brought to market as research and bioproduction reagents, mostly by Sigma-Aldrich. ProdiGene struck agreements with Sigma to distribute ProdiGene's corn-produced aprotinin, trypsin, beta-glucuronidase (GUS), and avidin. Large Scale Biology and SIgma agreed that Sigma would distribute LSBC's tobacco-produced aprotinin. Sigma also agreed to distribute Ventria's rise-produced Lactoferrin and Lysozyme.

However, in late 2002, just as ProdiGene was ramping up production of trypsin for commercial launch it was discovered that volunteer plants (leftover from the prior harvest) of one of their GM corn products were harvested with the conventional soybean crop later planted in that field. ProdiGene was fined $250,000 and ordered by the USDA to pay over $3 million in cleanup costs. This raised a furor and set the pharming field back, dramatically. Many companies went bankrupt as companies faced difficulties getting permits for field trials and investors fled. In reaction, APHIS introduced more strict regulations for pharming field trials in the US in 2003. In 2005, Anheuser-Busch threatened to boycott rice grown in Missouri because of plans by Ventria Bioscience to grow pharm rice in the state. A compromise was reached, but Ventria withdrew its permit to plant in Missouri due to unrelated circumstances.

The industry has slowly recovered, by focusing on pharming in simple plants grown in bioreactors and on growing GM crops in greenhouses. Some companies and academic groups have continued with open-field trials of GM crops that produce drugs. In 2006 Dow AgroSciences received USDA approval to market a vaccine for poultry against Newcastle disease, produced in plant cell culture – the first plant-produced vaccine approved in the U.S.

Pharming in Mammals

Historical Development

Milk is presently the most mature system to produce recombinant proteins from transgenic organisms. Blood, egg white, seminal plasma, and urine are other theoretically possible systems, but all have drawbacks. Blood, for instance, as of 2012 cannot store high levels of stable recombinant proteins, and biologically active proteins in blood may alter the health of the animals. Expression in the milk of a mammal, such as a cow, sheep, or goat, is a common application, as milk production is plentiful and purification from milk is relatively easy. Hamsters and rabbits have also been used in preliminary studies because of their faster breeding.

One approach to this technology is the creation of a transgenic mammal that can produce the biopharmaceutical in its milk (or blood or urine). Once an animal is produced, typically using the pronuclear microinjection method, it becomes efficacious to use cloning technology to create additional offspring that carry the favorable modified genome. In February 2009 the US FDA granted marketing approval for the first drug to be produced in genetically modified livestock. The drug is called ATryn, which is antithrombin protein purified from the milk of genetically modified goats. Marketing permission was granted by the European Medicines Agency in August 2006.

Patentability Issues Regarding Pharming

As indicated above, some mammals typically used for food production (such as goats, sheep, pigs, and cows) have been modified to produce non-food products, a practice sometimes called pharming. Use of genetically modified goats has been approved by the FDA and EMA to produce ATryn, i.e. recombinant antithrombin, an anticoagulant protein drug. These products "produced by turning animals into drug-manufacturing 'machines' by genetically modifying them" are sometimes termed biopharmaceuticals.

The patentability of such biopharmaceuticals and their process of manufacture is uncertain. Probably, the biopharmaceuticals themselves so made are unpatentable, assuming that they are chemically identical to the preexisting drugs that they imitate. Several 19th century United States Supreme Court decisions hold that a previously known natural product manufactured by artificial means cannot be patented. An argument can be made for the patentability of the process for manufacturing a biopharmaceutical, however, because genetically modifying animals so that they will produce the drug is dissimilar to previous methods of manufacture; moreover, one Supreme Court decision seems to hold open that possibility.

On the other hand, it has been suggested that the recent Supreme Court decision in *Mayo v. Prometheus* may create a problem in that, in accordance with the ruling in that case, "it may be said that such and such genes manufacture this protein in the same way they always did in a mammal, they produce the same product, and the genetic modification technology used is conventional, so that the steps of the process 'add nothing to the laws of nature that is not already present. If the argument prevailed in court, the process would also be ineligible for patent protection. This issue has not yet been decided in the courts.

Pharming in Plants

Plant-made pharmaceuticals (PMPs), also referred to as pharming, is a sub-sector of the biotechnology industry that involves the process of genetically engineering plants so that they can produce certain types of therapeutically important proteins and associated molecules such as peptides and secondary metabolites. The proteins and molecules can then be harvested and used to produce pharmaceuticals.

Recently, several non-crop plants such as the duckweed *Lemna minor* or the moss *Physcomitrella patens* have shown to be useful for the production of biopharmaceuticals. These frugal organisms can be cultivated in bioreactors (as opposed to being grown in fields), secrete the transformed proteins into the growth medium and, thus, substantially reduce the burden of protein purification in preparing recombinant proteins for medical use. In addition, both species can be engineered to cause secretion of proteins with human patterns of glycosylation, an improvement over conventional plant gene-expression systems. Biolex Therapeutics developed a duckweed-based expression platform; it sold that business to Synthon and declared bankruptcy in 2012.

Additionally, an Israeli company, Protalix, has developed a method to produce therapeutics in cultured transgenic carrot or tobacco cells. Protalix and its partner, Pfizer, received FDA approval to market its drug, taliglucerase alfa (Elelyso), treatment for Gaucher's disease, in 2012.

Arabidopsis is often used as a model organism to study gene expression in plants, while actual production may be carried out in maize, rice, potatoes, tobacco, flax or safflower. The advantage of rice and flax is that they are self-pollinating, and thus gene flow issues are avoided. However, human error could still result in pharm crops entering the food supply. Using a minor crop such as safflower or tobacco, avoids the greater political pressures and risk to the food supply involved with using staple crops such as beans or rice.

Regulation

The regulation of genetic engineering concerns the approaches taken by governments to assess and manage the risks associated with the development and release of genetically modified crops. There are differences in the regulation of GM crops - including those used for pharming - between countries, with some of the most marked differences occurring between the USA and Europe. Regulation varies in a given country depending on the intended use of the products of the genetic engineering. For example, a crop not intended for food use is generally not reviewed by authorities responsible for food safety.

Controversy Over Pharming

There are controversies around GMOs generally on several levels, including whether making them is ethical, issues concerning intellectual property and market dynamics; environmental effects of GM crops; and GM crops' role in industrial agricultural more generally. There are also specific controversies around pharming.

Advantages

Plants do not carry pathogens that might be dangerous to human health. Additionally, on the level of pharmacologically active proteins, there are no proteins in plants that are similar to human proteins. On the other hand, plants are still sufficiently closely related to animals and humans that they are able to correctly process and configure both animal and human proteins. Their seeds and fruits also provide sterile packaging containers for the valuable therapeutics and guarantee a certain storage life.

Global demand for pharmaceuticals is at unprecedented levels. Expanding the existing microbial systems, although feasible for some therapeutic products, is not a satisfactory option on several grounds. Many proteins of interest are too complex to be made by microbial systems or by protein synthesis. These proteins are currently being produced in animal cell cultures, but the resulting product is often prohibitively expensive for many patients. For these reasons, science has been exploring other options for producing proteins of therapeutic value.

These pharmaceutical crops could become extremely beneficial in developing countries. The World Health Organization estimates that nearly 3 million people die each year from vaccine preventable disease, mostly in Africa. Diseases such as measles and hepatitis lead to deaths in countries where the people cannot afford the high costs of vaccines, but pharm crops could help solve this problem.

Disadvantages

While molecular farming is one application of genetic engineering, there are concerns that are unique to it. In the case of genetically modified (GM) foods, concerns focus on the safety of the food for human consumption. In response, it has been argued that the genes that enhance a crop in some way, such as drought resistance or pesticide resistance, are not believed to affect the food itself. Other GM foods in development, such as fruits designed to ripen faster or grow larger, are believed not to affect humans any differently from non-GM varieties.

In contrast, molecular farming is not intended for crops destined for the food chain. It produces plants that contain physiologically active compounds that accumulate in the plant's tissues. Considerable attention is focused, therefore, on the restraint and caution necessary to protect both consumer health and environmental biodiversity.

The fact that the plants are used to produce drugs alarms activists. They worry that once production begins, the altered plants might find their way into the food supply or cross-pollinate with conventional, non-GM crops. These concerns have historical validation from the ProdiGene incident, and from the StarLink incident, in which GMO corn accidentally ended up in commercial food products. Activists also are concerned about the power of business. According to the Canadian Food Inspection Agency, in a recent report, says that U.S. demand alone for biotech pharmaceuticals is expanding at 13 percent annually and to reach a market value of $28.6 billion in 2004. Pharming is expected to be worth $100 billion globally by 2020.

List of originators (Companies and Universities), Research Projects and Products

- Dow AgroSciences - poultry vaccine against Newcastle disease virus (first PMP to be approved for marketing by the USDA Center for Veterinary Biologics Dow never intended to market the vaccine. "'Dow Agrosciences used the animal vaccine as an example to completely run through the process. A new platform needs to be approved, which can be difficult when authorities get in contact with it for the first time', explains the plant physiologist Stefan Schillberg, head of the Molecular Biology Division at the Fraunhofer Institute for Molecular Biology and Applied Ecology Aachen."

- Fraunhofer Institute for Molecular Biology and Applied Ecology, with sites in Germany, the US, and Chile is the lead institute of the Pharma Planta consortium of 33 partner organizations from 12 European countries and South Africa, funded by the European Commission. Pharma Planta is developing systems for plant production of proteins in greenhouses in the European regulatory framework. It is collaborating on biosimilars with Plantform and PharmaPraxis.

- Genzyme - antithrombin III in goat milk.

- GTC Biotherapeutics - ATryn (recombinant human antithrombin) in goat milk.

- Icon Genetics produces therapeutics in transiently infected *Nicotiana benthamiana* (relative of tobacco) plants in greenhouses in Halle, Germany or in fields. First product is a vaccine for a cancer, non-Hodgkin's lymphoma.

- Iowa State University - immunogenic protein from *E. coli* bacteria in pollen-free corn as a potential vaccine against *E. coli* for animals and humans.

- Kentucky Bioprocessing took over Large Scale Biology's facilities in Owensboro, Kentucky, and offers contract biomanufacturing services in tobacco plants, grown in greenhouses or in open fields.

- Medicago Inc. - Pre-clinical trials of Influenza vaccine made in transiently infected *Nicotiana benthamiana* (relative of tobacco) plants in greenhouses Medicago has a system for pharming in alfalfa that their website says is "not suited for the production of vaccines".

- PharmaPraxis - Developing biosimilars in collaboration with PlantForm and Fraunhofer.

- Pharming - C1 inhibitor, human collagen 1, fibrinogen (with American Red Cross), and lactoferrin in cow milk The intellectual property behind the fibrinogen project was acquired from PPL Therapeutics when PPL went bankrupt in 2004.

- Phyton Biotech uses plant cell culture systems to manufacture active pharmaceutical ingredients based on taxanes, including paclitaxel and docetaxel.

- Planet Biotechnology - antibodies against Streptococcus mutans, antibodies against doxorubicin, and ICAM 1 receptor in tobacco.

- PlantForm Corporation - biosimilar trastuzumab in tobacco - It is developing biosimilars in collaboration with PharmaPraxis and Fraunhofer.

- ProdiGene - was developing several proteins, including aprotinin, trypsin and a veterinary TGE vaccine in corn. Was in process of launching trypsin product in 2002 when later that year its field test crops contaminated conventional crops. Unable to pay the $3 million cost of the cleanup, it was purchased by International Oilseed Distributors in 2003 International Oilseed Distributors is controlled by Harry H. Stine, who owns one of the biggest soybeans genetics companies in the US. ProdiGene's maize-produced trypsin, with the trademark TrypZean is currently sold by Sigma-Aldritch as a research reagent.

- Syngenta - Beta carotene in rice (this is "Golden rice 2"), which Syngenta has donated to the Golden Rice Project.

- University of Arizona - Hepatitis C vaccine in potatoes.

- Ventria Bioscience - lactoferrin and lysozyme in rice.

- Washington State University - lactoferrin and lysozyme in barley.

- European COST Action on Molecular Farming - COST Action FA0804 on Molecular Farming provides a pan-European coordination centre, connecting academic and government institutions and companies from 23 countries. The aim of the Action is to advance the field by encouraging scientific interactions, providing expert opinion and encouraging commercial development of new products. The COST Action also provides grants allowing young scientists to visit participating laboratories across Europe for scientific training.

- Mapp Biopharmaceutical in San Diego, California was reported in August 2014 to be developing ZMapp, an experimental cure for the deadly Ebola virus disease. Two Americans who had been infected in Liberia were reported to be improving with the drug. ZMapp was made using antibodies produced by GM tobacco plants.

Projects known to be abandoned:

- Agragen, in collaboration with University of Alberta - docosahexaenoic acid and human serum albumin in flax.

- Chlorogen, Inc. - cholera, anthrax, and plague vaccines, albumin, interferon for liver diseases including hepatitis C, elastin, 4HB, and insulin-like growth factor in tobacco chloroplasts. Went out of business in 2007.

- Dow Chemical Company made a deal with Sunol Molecular in 2003 to develop antibodies against tissue factor in plants and in mammalian cell culture and to compare them. In 2005 Sunol sold all its tissue factor antagonists to Tanox, which in turn was bought by Genentech in 2003. Genentech licensed the tissue factor program to Altor in 2008 Altor is itself a spinout from Sunol. The product under development, ALT-836, formerly known as TNX-832 and Sunol-cH36, is the not the plant-produced antibody, but rather is mammalian antibody, more specifically, a chimeric antibody produced in a hybridoma.

- Epicyte - spermicidal antibodies in corn Epicyte was purchased by Biolex in 2004 at which time Epicyte's portfolio was described as "focused on the discovery and development of human monoclonal antibody products as treatments for a wide range of infectious and inflammatory diseases."

- Large Scale Biology Corporation (LSBC) (*bankrupt*) - used Tobacco mosaic virus to develop reagents and patient-specific vaccines for Non-Hodgkin's lymphoma, Papillomavirus vaccine, parvovirus vaccine, alpha galactosidase for Fabry disease, lysosomal acid lipase, aprotinin, interferon Alpha 2a and 2b, G-CSF, and Hepatitis B vaccine antigens in tobacco. In 2004, LSBC announced an agreement with Sigma-Aldritch under which LSB would produce recombinant aprotinin in plants of the tobacco family and Sigma-Aldrich would commercially distribute LSBC's recombinant product to its customers in the R&D, cell culture and manufacturing markets. As of October 2012 SIgma still has the protein in stock.

- Meristem Therapeutics - Lipase, lactoferrin, plasma proteins, collagen, antibodies (IgA, IgM), allergens and protease inhibitors in tobacco. Liquidated in 2008.

- Novoplant GmgH - therapeutic proteins in tobacco and feed peas. Conducted field trials in US of feed peas for pigs that produced anti-bacterial antibodies. Former CSO is now with another company; appears that Novoplant is out of business.

- Monsanto Company - abandoned development of pharmaceutical producing corn.

- PPL Therapeutics - Alpha 1-antitrypsin for cystic fibrosis and emphysema in sheep milk. This is the company that created Dolly the Sheep, the first cloned animal. Went bankrupt in 2004. Assets were acquired by Pharming and an investment group including University of Pittsburgh Medical Center.

- SemBioSys - insulin in safflower. In May 2012, SemBioSys terminated its operations.

Regulation of Genetically Modified Organisms

Regulations regarding the release of genetically modified organisms (GMOs) outside the laboratory varies widely by country. Countries such as the United States, Canada, Lebanon and Egypt use *substantial equivalence* as the starting point when assessing safety, while many countries such as those in the European Union, Brazil and China authorize GMO cultivation on a case-by-case basis. Many countries allow the import of GM food with authorization, but either do not allow its cultivation (Russia, Norway, Israel) or have provisions for cultivation, but no GM products are yet produced (Japan, South Korea). Most countries that do not allow for GMO cultivation do permit research.

One of the key issues concerning regulators is whether GM products should be labeled. Labeling of GMO products in the marketplace is required in 64 countries. Labeling can be mandatory up to a threshold GM content level (which varies between countries) or voluntary. A study investigating voluntary labeling in South Africa found that 31% of products labeled as GMO-free had a GM content above 1.0%. In Canada and the USA labeling of GM food is voluntary, while in Europe all food (including processed food) or feed which contains greater than 0.9% of approved GMOs must be labelled.

There is a scientific consensus that currently available food derived from GM crops poses no greater risk to human health than conventional food, but that each GM food needs to be tested on a case-by-case basis before introduction. Nonetheless, members of the public are much less likely than scientists to perceive GM foods as safe. The legal and regulatory status of GM foods varies by country, with some nations banning or restricting them, and others permitting them with widely differing degrees of regulation.

There is no evidence to support the idea that the consumption of approved GM food has a detrimental effect on human health. Some scientists and advocacy groups, such as Greenpeace and World Wildlife Fund, have however called for additional and more rigorous testing for GM food.

History

The development of a regulatory framework concerning genetic engineering began in 1975, at Asilomar, California. The first use of Recombinant DNA (rDNA) technology had just been successfully accomplished by Stanley Cohen and Herbert Boyer two years previously and the scientific community recognized that as well as benefits this technology could also pose some risks. The Asilomar meeting recommended a set of guidelines regarding the cautious use of recombinant technology and any products resulting from that technology. The Asilomar recommendations were voluntary, but in 1976 the US National Institute of Health (NIH) formed a rDNA advisory committee. This was followed by other regulatory offices (the United States Department of Agriculture (USDA), Environmental Protection Agency (EPA) and Food and Drug Administration (FDA)), effectively making all rDNA research tightly regulated in the USA. In 1982 the Organization for Economic Co-operation and Development (OECD) released a report into the potential hazards of releasing genetically modified organisms into the environment as the first transgenic plants were being developed. As the technology improved and genetically modified organisms moved from model organisms to potential commercial products the USA established a committee at the Office of Science and Technology (OSTP) to develop mechanisms to regulate the developing technology. In 1986 the OSTP assigned regulatory approval of genetically modified plants in the US to the USDA, FDA and EPA.

The basic concepts for the safety assessment of foods derived from GMOs have been developed in close collaboration under the auspices of the Organisation for Economic Co-operation and Development (OECD) and the United Nations' World Health Organisation (WHO) and Food and Agricultural Organisation (FAO). A first joint FAO/WHO consultation in 1990 resulted in the publication of the report 'Strategies for Assessing the Safety of Foods Produced by Biotechnology' in 1991. Building on that, an international consensus was reached by the OECD's Group of National Experts on Safety in Biotechnology, for assessing biotechnology in general, including field testing GM crops. That Group met again in Bergen, Norway in 1992 and reached consensus on principles for evaluating the safety of GM food; its report, 'The safety evaluation of foods derived by modern technology – concepts and principles' was published in 1993. That report recommends conducting the safety assessment of a GM food on a case-by-case basis through comparison to an existing food with a long history of safe use. This basic concept has been refined in subsequent workshops and consultations organized by the OECD, WHO, and FAO, and the OECD in particular has taken the lead in acquiring data and developing standards for conventional foods to be used in assessing substantial equivalence. In 2003 the Codex Alimentarius Commission of the FAO/WHO adopted a set of "Principles and Guidelines on foods derived from biotechnology" to help countries coordinate and standardize regulation of GM food to help ensure public safety and facilitate international trade and updated its guidelines for import and export of food in 2004.

Substantial Equivalence

"Substantial equivalence" is a starting point for the safety assessment for GM foods that is widely used by national and international agencies - including the Canadian Food Inspection Agency,

Japan's Ministry of Health and Welfare and the U.S. Food and Drug Administration, the United Nation's Food and Agriculture Organization, the World Health Organization and the OECD.

A quote from FAO, one of the agencies that developed the concept, is useful for defining it: "Substantial equivalence embodies the concept that if a new food or food component is found to be substantially equivalent to an existing food or food component, it can be treated in the same manner with respect to safety (i.e., the food or food component can be concluded to be as safe as the conventional food or food component)". The concept of substantial equivalence also recognises the fact that existing foods often contain toxic components (usually called antinutrients) and are still able to be consumed safely - in practice there is some tolerable chemical risk taken with all foods, so a comparative method for assessing safety needs to be adopted. For instance, potatoes and tomatoes can contain toxic levels of respectively, solanine and alpha-tomatine alkaloids.

To decide if a modified product is substantially equivalent, the product is tested by the manufacturer for unexpected changes in a limited set of components such as toxins, nutrients, or allergens that are present in the unmodified food. The manufacturer's data is then assessed by a regulatory agency, such as the U.S. Food and Drug Administration. That data, along with data on the genetic modification itself and resulting proteins (or lack of protein), is submitted to regulators. If regulators determine that the submitted data show no significant difference between the modified and unmodified products, then the regulators will generally not require further food safety testing. However, if the product has no natural equivalent, or shows significant differences from the unmodified food, or for other reasons that regulators may have (for instance, if a gene produces a protein that had not been a food component before), the regulators may require that further safety testing be carried out.

A 2003 review in *Trends in Biotechnology* identified seven main parts of a standard safety test:

1. Study of the introduced DNA and the new proteins or metabolites that it produces;

2. Analysis of the chemical composition of the relevant plant parts, measuring nutrients, anti-nutrients as well as any natural toxins or known allergens;

3. Assess the risk of gene transfer from the food to microorganisms in the human gut;

4. Study the possibility that any new components in the food might be allergens;

5. Estimate how much of a normal diet the food will make up;

6. Estimate any toxicological or nutritional problems revealed by this data in light of data on equivalent foods;

7. Additional animal toxicity tests if there is the possibility that the food might pose a risk.

There has been discussion about applying new biochemical concepts and methods in evaluating substantial equivalence, such as metabolic profiling and protein profiling. These concepts refer, respectively, to the complete measured biochemical spectrum (total fingerprint) of compounds (metabolites) or of proteins present in a food or crop. The goal would be to compare overall the biochemical profile of a new food to an existing food to see if the new food's profile falls within the range of natural variation already exhibited by the profile of existing foods or crops. However, these techniques are not considered sufficiently evaluated, and standards have not yet been developed, to apply them.

There are controversies over the definition and application of substantial equivalence.

By Continent

Africa

In 2010, after nine years of talks, the Common Market for Eastern and Southern Africa (COMESA) produced a draft policy on GM technology, which was sent to all 19 national governments for consultation in September 2010. Under the proposed policy, new GM crops would be scientifically assessed by COMESA. If the GM crop was deemed safe for the environmental and human health, permission would be granted for the crop to be grown in all 19 member countries, although the final decision would be left to each individual country.

In 2012, South Africa was the major commercial grower of genetically modified crops in Africa, with smaller amounts grown in Burkina Faso (maize), Egypt (cotton) and Sudan (cotton). Kenya passed laws in 2011, and Ghana and Nigeria passed laws in 2012 which allowed the production and importation of GM crops. By 2013 Cameroon, Malawi and Uganda had approved trials of genetically altered crops. A study investigating voluntary labeling in South Africa found that 31% of products labeled GMO-free had a GM content above 1.0%. 2011 studies for Uganda showed that transgenic bananas had a high potential to reduce rural poverty but that urban consumers with a relatively higher income might reject the introduction.

In 2002, Zambia cut off the flow of genetically modified food (mostly maize) from UN's World Food Programme on the basis of the Cartagena Protocol. This left the population without food aid during a famine. In December 2005 the Zambian government changed its mind in the face of further famine and allowed the importation of GM maize. However, the Zambian Minister for Agriculture Mundia Sikatana insisted in 2006, that the ban on genetically modified maize remained, saying "We do not want GM (genetically modified) foods and our hope is that all of us can continue to produce non-GM foods."

Asia

India and China are the two largest producers of genetically modified products in Asia. India currently only grows GM cotton, while China produces GM varieties of cotton, poplar, petunia, tomato, papaya and sweet pepper. Cost of enforcement of regulations in India are generally higher, possibly due to the greater influence farmers and small seed firms have on policy makers, while the enforcement of regulations was more effective in China. Other Asian countries that grew GM crops in 2011 were Pakistan, the Philippines and Myanmar. GM crops were approved for commercialisation in Bangladesh in 2013 and in Vietnam and Indonesia in 2014.

China

GM crops in China go through three phases of field trials (pilot field testing, environmental release testing, and preproduction testing) before they are submitted to the Office of Agricultural Genetic Engineering Biosafety Administration (OAGEBA) for assessment. Producers must apply to OAG-

EBA at each stage of the field tests. The Chinese Ministry of Science and Technology developed the first biosafety regulations for GM products in 1993 and they were updated in 2001. The 75 member National Biosafety Committee evaluates all applications, although OAGEBA has the final decision. Most of the National Biosafety Committee are involved in biotechnology leading to criticisms that they do not represent a wide enough range of public concerns.

India

The release of transgenic crops in India is governed by the Indian Environment Protection Act, which was enacted in 1986. The Institutional Biosafety Committee (IBSC), Review Committee on Genetic Manipulation (RCGM) and Genetic Engineering Approval Committee (GEAC) all review any genetically modified organism to be released, with transgenic crops also needing permission from the Ministry of Agriculture. India regulators cleared the Bt brinjal, a genetically modified eggplant, for commercialisation in October 2009. Following opposition from some scientists, farmers and environmental groups a moratorium was imposed on its release in February 2010.

Official Reports on GMO

There have been four official reports on GMO in India till August 2013 :

1. The 'Jairam Ramesh Report' - February 2010, imposing an indefinite moratorium on Bt Brinjal

2. The Sopory Committee Report - August 2012

3. The Parliamentary Standing Committee (PSC) Report on GM crops - August 2012

4. Final Report of The Technical Expert Committee established by Supreme Court - July 2013

Japan

Two laws regulate food safety and food quality in Japan, the Food Sanitation Law passed in 1947 and the Law Concerning Standardization and Proper Labeling of Agricultural and Forestry Products passed in 1950. The Food Sanitation Law has been amended and updated many times; an amendment dealing with pre-market approval and labeling of GMOs was passed in 2000 and came into effect in 2001. Japan passed laws to implement the Cartagena Protocol on Biosafety in September 2003 which came into effect in February 2004 - the Law Concerning the Conservation and Sustainable Use of Biological Diversity through Regulations on the Use of Living Modified Organisms (Law No. 97 of 2003).

Authority for approvals for various uses of genetically modified organisms is divided in Japan. The Ministry of the Environment has final approval for all uses of GMOs, but crops for commercial use and live vaccines for animals first go through the Ministry of Agriculture, Forestry and Fisheries; viruses for gene therapy and other medical applications first go through the Ministry of Health, Labor and Welfare; field trials of GM crops and recominant DNA used in biotechnology research first goes through the Ministry of Education, Culture, Sports, Science and Technology; and uses in the process of production of industrial enzymes, etc. goes through the Ministry of Economy, Trade and Industry.

Japan has not approved any commodity GM crops to be grown in Japan, but does allow import of agricultural products made from GM crops and food made of imported GM ingredients. Japan does however allow cultivation of GM flowers (e.g. Blue roses).

GM foods must undergo a safety assessment prior to being awarded certification for distribution to the domestic market. The Food Safety Commission (FSC) performs food and feed safety risk assessments.

Certain GM food must be labeled, but this is limited to designated genetically modified agricultural products, which are soybean, corn, potato, rapeseed, cottonseed, alfalfa and beet, and is limited to 32 processed foods which contain soybean, corn and potato, alfalfa and beet, in which recombinant DNA or the resulting protein still exists even after processing. However, processed food in which recombinant DNA or protein is dissolved in or removed during processing, such as soy sauce, soybean oil, corn flakes, millet jelly, corn oil, rapeseed oil, cottonseed oil, and others, do not have to be labeled.

Japan does not require traceability, and allows negative labeling ("GMO-free" and the like).

Philippines

The Philippines bans all GMOs recently overturning existing Department of Agriculture regulations. A petition filed on May 17, 2013 by environmental group Greenpeace Southeast Asia and farmer-scientist coalition Masipag (Magsasaka at Siyentipiko sa Pagpapaunlad ng Agrikultura) asked the appellate court to stop the planting of Bt eggplant in test fields, saying the impacts of such an undertaking to the environment, native crops and human health are still unknown. The Court of Appeals granted the petition, citing the precautionary principle stating "when human activities may lead to threats of serious and irreversible damage to the environment that is scientifically plausible but uncertain, actions shall be taken to avoid or diminish the threat." Respondents filed a motion for reconsideration in June 2013 and on September 20, 2013 the Court of Appeals chose to uphold their May decision saying the bt talong field trials violate the people's constitutional right to a "balanced and healthful ecology." The Supreme Court on Tuesday, December 8, 2015 permanently stopped the field testing for Bt (Bacillus thuringiensis) talong (eggplant), upholding the decision of the Court of Appeals which stopped the field trials for the genetically modified eggplant. The Philippines Supreme Court also took the unprecedented step and invalidated the Department of Agriculture administrative order allowing the field testing, propagation and commercialization, and importation of GMOs.

Europe

Until the 1990s, Europe's regulation was less strict than in the United States, one turning point being cited as the export of the United States' first GM-containing soy harvest in 1996. The GM soy made up about 2% of the total harvest at the time, and Eurocommerce and European food retailers required that it be separated. In 1998, the use of MON810, a Bt expressing maize conferring resistance to the European corn borer, was approved for commercial cultivation in Europe. Shortly thereafter, the EU enacted a *de facto* moratorium on new approvals of GMOs pending new regulatory laws passed in 2003.

Those new laws provided the European Union (EU) with possibly the most stringent GMO regulations in the world. All GMOs, along with irradiated food, are considered "new food" and subject to extensive, case-by-case, science based food evaluation by the European Food Safety Authority (EFSA). The criteria for authorization fall in four broad categories: "safety," "freedom of choice," "labelling," and "traceability." The EFSA reports to the European Commission who then draft a proposal for granting or refusing the authorisation. This proposal is submitted to the Section on GM Food and Feed of the Standing Committee on the Food Chain and Animal Health and if accepted it will be adopted by the EC or passed on to the Council of Agricultural Ministers. Once in the Council it has three months to reach a qualified majority for or against the proposal, if no majority is reached the proposal is passed back to the EC who will then adopt the proposal. However, even after authorization, individual EU member states can ban individual varieties under a 'safeguard clause' if there are "justifiable reasons" that the variety may cause harm to humans or the environment. The member state must then supply sufficient evidence that this is the case. The Commission is obliged to investigate these cases and either overturn the original registrations or request the country to withdraw its temporary restriction. The laws of the EU also stipulated that member nations establish coexistence regulations. In many cases national coexistence regulations include minimum distances between fields of GM crops and non-GM crops. The distances for GM maize from non-GM maize for the six largest biotechnology countries are; France: 50 meters, Britain: 110 meters for grain maize and 80 for silage maize, Netherlands: 25 meters in general and 250 for organic or GM-free fields, Sweden: 15–50 meters, Finland: data not available, and Germany: 150 meters and 300 from organic fields. Larger minimum distance requirements discriminate against adoption of GM crops by smaller farms.

In 2006, the World Trade Organization concluded that the EU moratorium, which had been in effect from 1998 to 2004, had violated international trade rules. The moratorium had not affected previously approved crops. The only crop authorised for cultivation before the moratorium was Monsanto's MON 810. The next approval for cultivation was the Amflora potato for industrial applications in 2010 which was grown in Germany, Sweden and the Czech Republic that year.

The slow pace of approval has been criticized as endangering European food safety although as of 2012, the EU has authorized the use of 48 genetically modified organisms. Most of these were for use in animal feed (it was reported in 2012 that the EU imports about 30 million tons a year of GM crops for animal consumption.), food or food additives. 26 of these were varieties of maize. In July 2012 the EU gave approval for an Irish trial cultivation of potatoes resistant to the blight that caused the Great Irish Famine.

The safeguard clause mentioned above has been applied by many member states in various circumstances, and in April 2011 there were 22 active bans in place across six member states: Austria, France, Germany, Luxembourg, Greece, and Hungary. However, on review many of these have been considered scientifically unjustified.

- In January 2005, the Hungarian government announced a ban on importing and planting of genetic modified maize seeds, which was subsequently authorized by the EU.

- In February 2008 the French government used the safeguard clause to ban the cultivation of MON810 after Senator Jean-François Le Grand, chairman of a committee set up to evaluate biotechnology, said there were "serious doubts" about the safety of the product (although this ban was declared illegal in 2011 by the European Court of Justice and the

French Conseil d'État). The French farm ministry reinstated the ban in 2012, but this was rejected by the EFSA.

- In 2009 German Federal Minister Ilse Aigner announced an immediate halt to cultivation and marketing of MON810 maize under the safeguard clause.

- In March 2010, Bulgaria imposed a complete ban on genetically modified crop growing either commercially or for trials. The cabinet of Boyko Borisov initially imposed a 5-year moratorium, but later extended it to a permanent ban after widespread public protests against the introduction of genetically modified crops in the country. And in recent years, France and several other European countries banned cultivation of Monsanto's MON-810 corn and similar genetically modified food crops.

- Since January 2013 Poland's government placed a ban on Monsanto's GM corn, MON 810 and has launched a communication campaign with farmers', announcing they will now be strictly monitoring farms for GM corn crops. Poland is the eighth EU member to ban the production of GMOs although they have been approved by European Food Safety Authority. Europe is not against the use of GM crops when it comes to laboratory research, they are working to regulate the field.

In 2012, the European Food Safety Authority (EFSA) Panel on Genetically Modified Organisms (GMO) released a "Scientific opinion addressing the safety assessment of plants developed through cisgenesis and intragenesis" in a response to a request from the European Commission. The opinion was, that while "the frequency of unintended changes may differ between breeding techniques and their occurrence cannot be predicted and needs to be assessed case by case," "similar hazards can be associated with cisgenic and conventionally bred plants, while novel hazards can be associated with intragenic and transgenic plants." In other words, cisgenic genetic engineering approaches should be considered similar in risk to conventional breeding approaches, each of which are less risky than transgenic approaches.

In 2014 a panel of experts set up by the UK Biotechnology and Biological Sciences Research Council argued that "A regulatory system based on the characteristics of a novel crop, by whatever method it has been produced, would provide a more effective and robust regulation than current EU processes , which consider new crop varieties differently depending on the method used to produce them." They said that new forms of "genome editing" allow targeting specific sites and making precise changes in the DNA of crops. In the future it would become increasingly difficult if not impossible to tell which method has been used (conventional breeding or genetic engineering) to produce a novel crop. They proposed that existing EU regulatory system should be replaced with a more logical system like that used for new medicines.

In 2015 Germany, Poland, France, Scotland and several other member states opted out of cultivating GMO crops in their territory.

Labeling and Traceability

The regulations concerning the import and sale of GMOs for human and animal consumption grown outside the EU involve providing freedom of choice to the farmers and consumers. All food (including processed food) or feed which contains greater than 0.9% of approved GMOs must be labelled. Twice GMOs unapproved by the EC have arrived in the EU and been forced to return to their port of origin. The first was in 2006 when a shipment of rice from America containing an experimental GMO variety (LLRice601) not meant for commercialisation arrived at Rotterdam. The second in 2009 when trace amounts of a GMO maize approved in the US were found in a "non-GM" soy flour cargo.

The coexistence has raised significant concern in many European countries and so EU law also requires that all GM food be traceable to its origin, and that all food with GM content greater than 0.9% be labelled. Due to high demand from European consumers for freedom of choice between GM and non-GM foods. EU regulations require measures to avoid mixing of foods and feed produced from GM crops and conventional or organic crops, which can be done via isolation distances or biological containment strategies. (Unlike the US, European countries require labeling of GM food.) European research programs such as Co-Extra, Transcontainer, and SIGMEA are investigating appropriate tools and rules for traceability. The OECD has introduced a "unique identifier" which is given to any GMO when it is approved, which must be forwarded at every stage of processing. Such measures are generally not used in North America because they are very costly and the industry admits of no safety-related reasons to employ them. The EC has issued guidelines to allow the co-existence of GM and non-GM crops through buffer zones (where no GM crops are grown). These are regulated by individual countries and vary from 15 meters in Sweden to 800 meters in Luxembourg. All food (including processed food) or feed which contains greater than 0.9% of approved GMOs must be labelled.

North America

As of 2002 the United States, Canada, and Mexico did not require labeling of genetically modified foods.

Canada

Mainland Canada is one of the world's largest producers of GM canola and also grows GM maize, soybean and sugarbeet. Health Canada, under the Food and Drugs Act, and the Canadian Food Inspection Agency are responsible for evaluating the safety and nutritional value of genetically modified foods. Environmental assessments of biotechnology-derived plants are carried out by the CFIA's Plant Biosafety Office (PBO). The Canadian regulatory system is based on whether a product has novel features regardless of method of origin. In other words, a product is regulated as GM if it carries some trait not previously found in the species whether it was generated using traditional breeding methods (e.g. selective breeding, cell fusion, mutation breeding) or genetic engineering. Canadian law requires that manufacturers and importers submit detailed scientific data to Health Canada for safety assessments for approval. This data includes: information on how the GM plant was developed; nucleic acid data that characterizes the genetic change; composition and nutritional data of the novel food compared to the original non-modified food' potential for new toxins; and potential for being an allergen. A decision is then made whether to approve the product for release along with any restrictions or requirements. Labeling of foods as products of Genetic Engineering or not products of Genetic Engineering is voluntary. The Canadian regulations were reviewed by the Canadian Biotechnology Advisory Committee between 1999 and 2003, with the conclusion that the current level of regulation was satisfactory. The committee was accused by environmental and citizen groups of not representing the full spectrum of public interests by only having one member of the board of 20 representing non-governmental organisations and for being too closely aligned to industry groups.

Mexico

In February 2005, after consulting the Mexican Academy of Sciences, Mexico's senate passed a law allowing to plant and sell genetically modified cotton and soybean. The law requires all genetically modified products to be labelled according to guidelines issued by the Mexican Ministry of Health. In 2009, the government enacted statutory provisions for the regulation of genetically modified maize. Mexico is the center of diversity for maize and concerns had been raised about the impact genetically modified maize could have on local strains. In 2013, a federal judge ordered Mexico's SAGARPA (Secretaría de Agricultura, Ganadería, Desarrollo Rural, Pesca, y Alimentación), which is Mexico's Secretary of Agriculture, and SEMARNAT (Secretaría de Medio Ambiente y Recursos Naturales), equivalent of the EPA, to temporarily halt any new GMO corn permits, accepting a lawsuit brought by opponents of the crop.

United States

Federal Regulation

The USA is the largest commercial grower of genetically modified crops in the world.

United States regulatory policy is governed by the Coordinated Framework for Regulation of Biotechnology This regulatory policy framework that was developed under the Presidency of Ronald Reagan to ensure safety of the public and to ensure the continuing development of the fledgling biotechnology industry without overly burdensome regulation. The policy as it developed had three tenets: "(1) U.S. policy would focus on the product of genetic modification (GM) techniques, not the process itself, (2) only regulation grounded in verifiable scientific risks would be tolerated, and (3) GM products are on a continuum with existing products and, therefore, existing statutes are sufficient to review the products." In 2015 the Obama administration announced that it would update the way the government regulated genetically modified crops.

For a genetically modified organism to be approved for release, it must be assessed under the Plant Protection Act by the Animal and Plant Health Inspection Service (APHIS) agency within the US Department of Agriculture (USDA) and may also be assessed by the Food and Drug Administration (FDA) and the Environmental protection agency (EPA), depending on the intended use of the organism. The USDA evaluates the plant's potential to become a weed. The FDA has a voluntary consultation process with the developers of genetically engineered plants. The Federal Food, Drug, and Cosmetic Act, which outlines FDA's responsibilities, does not require pre-market clearance of food, including genetically modified food plants. The EPA regulates genetically modified plants with pesticide properties, as well as agrochemical residues. Most genetically modified plants are reviewed by at least two of the agencies, with many subject to all three. Within the organization are departments that regulate different areas of GM food including, the Center for Food Safety and Applied Nutrition (CFSAN,) and the Center for Biologics Evaluation and Research (CBER). As of 2008, all developers of genetically modified crops in the US had made use of the voluntary process. Final approval can still be denied by individual counties within each state. In 2004, Mendocino

County, California became the first county to impose a ban on the "Propagation, Cultivation, Raising, and Growing of Genetically Modified Organisms", the measure passing with a 57% majority. In May, 2014 Jackson and Josephine Counties in Southern Oregon passed initiatives similar to that passed by Mendocino County; both passing by 2 to 1 margins.

Several laws govern the US regulatory agencies. These laws are statutes the agencies review when determining the safety of a particular GM food. These laws include:

- The Federal Insecticide, Fungicide, and Rodenticide Act (FIFRA) (EPA);

- The Toxic Substances Control Act (TSCA) (EPA);

- The Federal Food, Drug, and Cosmetic Act (FFDCA) (FDA and EPA);

- The Plant Protection Act (PPA) (USDA);

- The Virus-Serum-Toxin Act (VSTA) (USDA);

- The Public Health Service Act (PHSA)(FDA);

- The Dietary Supplement Health and Education Act (DSHEA) (FDA)

- The Meat Inspection Act (MIA)(USDA);

- The Poultry Products Inspection Act (PPIA) (USDA);

- The Egg Products Inspection Act (EPIA) (USDA); and

- The National Environmental Protection Act (NEPA).

State Regulation

Several states have passed regulations concerning labelling of GM food; Connecticut passed a GMO labeling bill in May 2013, but the bill will only be triggered after four other states enact similar legislation. On January 9, 2014, Maine's governor signed a bill requiring labeling for foods made with GMO's, with a similar triggering mechanism as Connecticut's bill. In May 2014 Vermont passed a law requiring labeling of food containing ingredients derived from genetically modified organisms. A federal judge ruled Maui's GMO ban invalid.

South America

Brazil and Argentina are the 2nd and 3rd largest producers of GM food behind the USA.

The Argentine government was one of the first to accept GM food. Assessment of GM products for release is provided by the National Agricultural Biotechnology Advisory Committee (environmental impact), the National Service of Health and Agrifood Quality (food safety) and the National Agribusiness Direction (effect on trade), with the final decision made by the Secretariat of Agriculture, Livestock, Fishery and Food. The government is looking to tighten the current law which allows farmers to keep seed without paying royalties in a bid to encourage more private investment.

In Brazil the National Biosafety Technical Commission is responsible for assessing environmental and food safety and prepares guidelines for transport, importation and field experiments involving GM products. The Council of Ministers evaluates the commercial and economical issues with release. The National Biosafety Technical Commission has 27 members and includes 12 scientists, 9 ministerial representatives and 6 other specialists.

Honduras, Costa Rica, Colombia, Bolivia, Paraguay, Chile, and Uruguay also allow GM crops to be grown.

Venezuela banned genetically modified seeds in 2004, in 2008, Ecuador prohibited genetically engineered crops and seeds in its 2008 Constitution, approved by 64% of the population in a referendum (although Ecuadorian President Rafael Correa said in 2012 that this was "a mistake". Peru has banned transgenic crops.

Oceania

Malaysia, New Zealand, and Australia require labeling so consumers can exercise choice between foods that have genetically modified, conventional or organic origins.

Australia

Genetic engineering in Australia was originally (since 1987) overseen by the Genetic Manipulation Advisory Committee, before the Office of the Gene Technology Regulator (OGTR) and Food Standards Australia New Zealand took over in 2001. The OTGR is a Commonwealth Government Authority within the Department of Health and Ageing and reports directly to Parliament through a Ministerial Council on Gene Technology and has legislative powers. It was established as part of the Gene Technology Act 2003 and operates according to the Gene Technology Regulations 2001. The OGTR reports directly to Parliament through a Ministerial Council on Gene Technology and has legislative powers. The OGTR decides on license applications for the release of all genetically modified organisms, while regulation is provided by the Therapeutic Goods Administration for GM medicines or Food Standards Australia New Zealand for GM food. The individual state governments are then able to assess the impact of release on markets and trade and apply further legislation to control approved genetically modified products.

Genetically modified cotton, canola, and carnations are grown in Australia. Genetically modified cotton has been grown commercially in New South Wales and Queensland since 1996. GM canola was approved in 2003 and was first grown in 2008 and was first approved in Western Australia in 2010.

In 2011 genetically modified plants were grown in all states except South Australia and Tasmania, who have extended their moratoriums until 2019 and 2014. The Queensland and Northern Territory Governments have not implemented any further legislation beyond the national level, but several other states placed bans on planting certain GM crops. In 2007 the New South Wales government extended a blanket moratorium on GM food crops until 2011, but allowed groups to apply for exemptions. New South Wales approved GM Canola for commercial cultivation in 2008, while the Victorian government let the moratorium on GM Canola expire in 2007. Western Australia passed the Genetically Modified Crops Free Areas Act in 2003 and was declared a GM free area in 2004. In 2008 an exception was made for the commercial cultivation of GM cotton in the

Ord River Irrigation Areas. Trials of GM canola were carried out in 2003 and in 2010 the Western Australian government allowed the commercialisation of GM canola.

New Zealand

As of 2004 no genetically modified food was grown in New Zealand, and no medicines containing live genetically modified organisms have been approved for use. However, medicines manufactured using genetically modified organisms that do not contain live organisms have been approved for sale, and imported foods with genetically modified components are sold. In 2000 the Government appointed a Royal Commission to report on issues relating to genetically modified organisms (GMOs). The Report of the Royal Commission on Genetic Modification, released in July 2001, concluded that New Zealand should keep its options open with regard to genetic engineering and to proceed carefully in order to minimise and manage any risks. Field trials have been carried out with GM pine trees and brassicas. Food Standards Australia New Zealand (FSANZ) must approve any food produced from GM crops, or made using genetically engineered enzymes, before it can be marketed in Australia or New Zealand. FSANZ makes a list of such approvals available on its website.

Transgene

A transgene is a gene or genetic material that has been transferred naturally, or by any of a number of genetic engineering techniques from one organism to another. The introduction of a transgene (called "transgenesis") has the potential to change the phenotype of an organism.

In its most precise usage, the term *transgene* describes a segment of DNA containing a gene sequence that has been isolated from one organism and is introduced into a different organism. This non-native segment of DNA may either retain the ability to produce RNA or protein in the transgenic organism or alter the normal function of the transgenic organism's genetic code. In general, the DNA is incorporated into the organism's germ line. For example, in higher vertebrates this can be accomplished by injecting the foreign DNA into the nucleus of a fertilized ovum. This technique is routinely used to introduce human disease genes or other genes of interest into strains of laboratory mice to study the function or pathology involved with that particular gene.

The construction of a transgene requires the assembly of a few main parts. The transgene must contain a promoter, which is a regulatory sequence that will determine where and when the transgene is active, an exon, a protein coding sequence (usually derived from the cDNA for the protein of interest), and a stop sequence. These are typically combined in a bacterial plasmid and the coding sequences are typically chosen from transgenes with previously known functions.

Transgenic or genetically modified organisms, be they bacteria, viruses or fungi, serve all kinds of research purposes. Transgenic plants, insects, fish and mammals have been bred. Transgenic plants such as corn and soybean have replaced wild strains in agriculture in some countries (e.g. the United States). Transgene escape has been documented for GMO crops since 2001 with persistence and invasiveness. Transgenetic organisms pose ethical questions and may cause biosafety problems.

History

The idea of shaping an organism to fit a specific need isn't a new science; selective breeding of animals and plants started before recorded history. However, until the late 1900s farmers and scientist could breed new strains of a plant or organism only from closely related species, because the DNA had to be compatible for offspring to be able to reproduce another generation.

In the 1970 and 1980s, scientists passed this hurdle by inventing procedures for combining the DNA of two vastly different species with genetic engineering. The organisms produced by these procedures were termed transgenic. Transgenesis is the same as gene therapy in the sense that they both transform cells for a specific purpose. However, they are completely different in their purposes, as gene therapy aims to cure a defect in cells, and transgenesis seeks to produce a genetically modified organism by incorporating the specific transgene into every cell and changing the genome. Transgenesis will therefore change the germ cells, not only the somatic cells, in order to ensure that the transgenes are passed down to the offspring when the organisms reproduce. Transgenes alter the genome by blocking the function of a host gene; they can either replace the host gene with one that codes for a different protein, or introduce an additional gene.

In 1978, yeast cells were the first organisms to undergo gene transfer. Mouse cells were first transformed in 1979, followed by mouse embryos in 1980. Most of the very first transmutations were performed by microinjection of DNA directly into cells. Scientist were able to develop other methods to perform the transformations, such as incorporating transgenes into retroviruses and then infecting cells, using electroinfusion which takes advantage of an electric current to pass foreign DNA through the cell wall, biolistics which is the procedure of shooting DNA bullets into cells, and also delivering DNA into the egg that has just been fertilized.

The first transgenic animals were only intended for genetic research to study the specific function of a gene, and by 2003, thousands of genes had been studied.

Use in Plants

A Variety Of transgenic plants have been designed for agriculture to produce genetically modified crops, such as corn, soybean, rapeseed oil, cotton, rice and more. As of 2012, these GMO crops were planted on 170 million hectares globally.

Golden Rice

One example of a transgenic plant species is golden rice. In 1997, five million children developed xerophthalmia, a medical condition caused by vitamin A deficiency, in Southeast Asia alone. Of those children, a quarter million went blind. To combat this, scientists used biolistics to insert the daffodil phytoene synthase gene into Asia indigenous rice cultivars. The daffodil insertion increased the production ß-carotene. The product was a transgenic rice species rich in vitamin A, called golden rice. Little is known about the impact of golden rice on xerophthalmia because anti-GMO campaigns have prevented the full commercial release of golden rice into agricultural systems in need.

Transgene Escape

The escape of genetically-engineered plant genes via hybridization with wild relatives was first discussed and examined in Mexico and Europe in the mid-1990s. There is agreement that escape of transgenes is inevitable, even "some proof that it is happening". Up until 2008 there were few documented cases.

Corn

Corn sampled in 2000 from the Sierra Juarez, Oaxaca, Mexico contained a transgenic 35S promoter, while a large sample taken by a different method from the same region in 2003 and 2004 did not. A sample from another region from 2002 also did not, but directed samples taken in 2004 did, suggesting transgene persistence or re-introduction. A 2009 study found recombinant proteins in 3.1% and 1.8% of samples, most commonly in southeast Mexico. Seed and grain import from the United States could explain the frequency and distribution of transgenes in west-central Mexico, but not in the southeast. Also, 5.0% of corn seed lots in Mexican corn stocks expressed recombinant proteins despite the moratorium on GM crops.

Cotton

In 2011, transgenic cotton was found in Mexico among wild cotton, after 15 years of GMO cotton cultivation.

Rapeseed (Canola)

Transgenic rapeseed *Brassicus napus*, hybridized with a native Japanese species *Brassica rapa*, was found in Japan in 2011 after they had been identified 2006 in Québec, Canada. They were persistent over a 6-year study period, without herbicide selection pressure and despite hybridization with the wild form. This was the first report of the introgression—the stable incorporation of genes from one gene pool into another—of an herbicide resistance transgene from *Brassica napus* into the wild form gene pool.

Creeping Bentgrass

Transgenic creeping bentgrass, engineered to be glyphosate-tolerant as "one of the first wind-pollinated, perennial, and highly outcrossing transgenic crops", was planted in 2003 as part of a large (about 160 ha) field trial in central Oregon near Madras, Oregon. In 2004, its pollen was found to have reached wild growing bentgrass populations up to 14 kilometres away. Cross-pollinating *Agrostis gigantea* was even found at a distance of 21 kilometres. The grower, Scotts Company could not remove all genetically engineered plants, and in 2007, the U.S. Department of Agriculture fined Scotts $500,000 for noncompliance with regulations in 2007.

Risk Assessment

The long-term monitoring and controlling of a particular transgene has been shown not to be feasible. The European Food Safety Authority published a guidance for risk assessment in 2010.

Use in Mice

Genetically modified mice are the most common animal model for transgenic research. Transgenic mice are currently being used to study a variety of diseases including cancer, obesity, heart disease, arthritis, anxiety, and Parkinson's disease. The two most common types of genetically modified mice are knockout mice and oncomice. Knockout mice are a type of mouse model that uses transgenic insertion to disrupt an existing gene's expression. In order to create knockout mice, a transgene with the desired sequence is inserted into an isolated mouse blastocyst using electroporation. Then, homologous recombination occurs naturally within some cells, replacing the gene of interest with the designed transgene. Through this process, researchers were able to demonstrate that a transgene can be integrated into the genome of an animal, serve a specific function within the cell, and be passed down to future generations.

Oncomice are another genetically modified mouse species created by inserting transgenes that increase the animal's vulnerability to cancer. Cancer researchers utilize oncomice to study the profiles of different cancers in order to apply this knowledge to human studies.

Use in *Drosophila*

Multiple studies have been conducted concerning transgenesis in *Drosophila melanogaster*, the fruit fly. This organism has been a helpful genetic model for over 100 years, due to its well-understood developmental pattern. The transfer of transgenes into the *Drosophila* genome has been performed using various techniques, including P element, Cre-loxP, and ΦC31 insertion. The most practiced method used thus far to insert transgenes into the *Drosophila* genome utilizes P elements. The transposable P elements, also known as transposons, are segments of bacterial DNA that are translocated into the genome, without the presence of a complementary sequence in the host's genome. P elements are administered in pairs of two, which flank the DNA insertion region of interest. Additionally, P elements often consist of two plasmid components, one known as the P element transposase and the other, the P transposon backbone. The transposase plasmid portion drives the transposition of the P transposon backbone, containing the transgene of interest and often a marker, between the two terminal sites of the transposon. Success of this insertion results in the nonreversible addition of the transgene of interest into the genome. While this method has been proven effective, the insertion sites of the P elements are often uncontrollable, resulting in an unfavorable, random insertion of the transgene into the *Drosophila* genome.

To improve the location and precision of the transgenic process, an enzyme known as Cre has been introduced. Cre has proven to be a key element in a process known as recombination-mediated cassette exchange (RMCE). While it has shown to have a lower efficiency of transgenic transformation than the P element transposases, Cre greatly lessens the labor-intensive abundance of balancing random P insertions. Cre aids in the targeted transgenesis of the DNA gene segment of interest, as it supports the mapping of the transgene insertion sites, known as loxP sites. These sites, unlike P elements, can be specifically inserted to flank a chromosomal segment of interest, aiding in targeted transgenesis. The Cre transposase is important in the catalytic cleavage of the base pairs present at the carefully positioned loxP sites, permitting more specific insertions of the transgenic donor plasmid of interest.

To overcome the limitations and low yields that transposon-mediated and Cre-loxP transformation methods produce, the bacteriophage ΦC31 has recently been utilized. Recent breakthrough studies involve the microinjection of the bacteriophage ΦC31 integrase, which shows improved transgene insertion of large DNA fragments that are unable to be transposed by P elements alone.

This method involves the recombination between an attachment (attP) site in the phage and an attachment site in the bacterial host genome (attB). Compared to usual P element transgene insertion methods, ΦC31 integrates the entire transgene vector, including bacterial sequences and antibiotic resistance genes. Unfortunately, the presence of these additional insertions has been found to affect the level and reproducibility of transgene expression.

Future Potential

The study of application of transgenes is a rapidly growing area of molecular biology. In fact, it is predicted that in the next two decades, 300 000 lines of transgenic mice will be generated. Researchers have identified many applications for transgenes, particularly in the medical field. Scientists are focusing on the use of transgenes to study the function of the human genome in order to better understand disease, adapting animal organs for transplantation into humans, and the production of pharmaceutical products such as insulin, growth hormone, and blood anti-clotting factors from the milk of transgenic cows.

There are currently five thousand known genetic diseases, and the potential to treat these diseases using transgenic animals is, perhaps, one of the most promising applications of transgenes. There is a potential to use human gene therapy to replace a mutated gene with an unmutated copy of a transgene in order to treat the genetic disorder. This can be done through the use of Cre-Lox or knockout. Moreover, genetic disorders are being studied through the use of transgenic mice, pigs, rabbits, and rats. More recently, scientists have also begun using transgenic goats to study genetic disorders related to fertility.

Transgenes may soon be used for xenotransplantation from pig organs. Through the study of xeno-organ rejection, it was found that an acute rejection of the transplanted organ occurs upon the organs contact with blood from the recipient due to the recognition of foreign antibodies on endothelial cells of the transplanted organ. Scientists have identified the antigen in pigs that causes this reaction, and therefore are able to transplant the organ without immediate rejection by removal of the antigen. However, the antigen begins to be expressed later on, and rejection occurs. Therefore, further research is being conducted.

Transgenes are being used by manufactures to produce goods such as milk with high levels of proteins, silk from the milk of goats, and microorganisms that are capable of producing proteins that contain enzymes that increase the rate of industrial reactions. Agricultural applications aim to selectively breed animals for particular traits and animals that are resistant to diseases.

Ethical Controversy

Transgene use in humans is currently fraught with issues. Transformation of genes into human cells has not been perfected yet. The most famous example of this involved certain patients developing T-cell leukemia after being treated for X-linked severe combined immunodeficiency (X-SCID). This was attributed to the close proximity of the inserted gene to the LMO2 promoter, which controls the transcription of the LMO2 proto-oncogene. In common with most forms of genetic engineering, the use of transgenes for purposes other than to correct life-threatening genetic abnormalities is a major bioethical issue.

Transgenesis

Transgenesis is the process of introducing an exogenous gene—called a transgene—into a living organism so that the organism will exhibit a new property and transmit that property to its offspring. Transgenesis can be facilitated by liposomes, enzymes, plasmid vectors, viral vectors, pronuclear injection, protoplast fusion, and ballistic DNA injection. Transgenesis can occur in nature.

Transgenic Organisms are able to express foreign genes because the genetic code is similar for all organisms. This means that a specific DNA sequence will code for the same protein in all organisms. Due to this similarity in protein sequence, scientists can cut DNA at these common protein points and add other genes. An example of this is the "super mice" of the 1980s. These mice were able to produce the human protein tPA to treat blood clots.

Using Plasmids from Bacteria

The most common type of transgenesis research is done with bacteria and viruses which are able to replicate foreign DNA. The plasmid DNA is cut using restriction enzymes, while the DNA to be copied is also cut with the same restriction enzyme, producing complementary sticky-ends. This allows the foreign DNA to hybridise with the plasmid DNA and be sealed by DNA ligase enzyme, creating a genetic code not normally found in nature. Altered DNA is inserted into plasmids for replication.

Gene Transfer Technology

DNA Microinjection

The Desired gene construct is injected in the pronucleus of a reproductive cell using a glass needle around 0.5 to 5 micrometers in diameter. The manipulated cell is cultured in vitro to develop to a specific embryonic phase, is then transferred to a recipient female. DNA microinjection does not have a high success rate (roughly 2% of all injected subjects), even if the new DNA is incorporated in the genome, if it is not accepted by the germ-line the new traits will not appear in their offspring. If DNA is injected in multiple sites the chances of over-expression increase.

Retrovirus-mediated Gene Transfer

A retrovirus is a virus that carries its genetic material in the form of RNA rather than DNA. Retroviruses are used as vectors to transfer genetic material into the host cell. The result is a chimera, an organism consisting of tissues or parts of diverse genetic constitution. Chimeras are inbred for as many as 20 generations until homozygous genetic offspring are born.

Restriction Enzyme Mediated Integration

Restriction enzyme mediated integration (REMI) is a technique for integrating DNA (linearised plasmid) into the genome sites that have been generated by the same restriction enzyme used for the DNA linearisation. The plasmid integration occurs at the corresponding sites in the genome, often by regenerating the recognition sites by same the restriction enzyme used for plasmid linearisation.

Stem Cell Transgenesis

Multipotent Stem Cell Transgenesis

Multipotent stem cells can only differentiate into a limited number of therapeutically useful cell types, nevertheless their safety and relative lack of complexity to us have resulted in the vast majority of current personalized cellular therapeutics involving multipotent stem cells (typically mesenchymal stem cells from adipose tissue).

Pluripotent Stem Cell Transgenesis

Transgenic vectors can be delivered randomly, or targeted to a specific genomic location, such as a safe harbor. Scientists have performed research and technology development to provide the tools necessary to permit safe and effective pluripotent stem cell (PSC) transgenesis.

Totipotent Stem Cell Transgenesis

The manipulated gene construct is inserted into totipotent stem cells, cells which can develop into any specialized cell. Cells containing the desired DNA are incorporated into the host's embryo, resulting in a chimeric animal. Unlike the other two methods of injection which require live transgenic offspring for testing, embryonic cell transfer can be tested at the cell stage.

Applications

Pharming

Pharming, a portmanteau of "farming" and "pharmaceutical", refers to the use of genetic engineering to insert genes that code for useful pharmaceuticals into host animals or plants that would otherwise not express those genes. Pharming has gained application in biotechnology since the development of transgenic "super mice" in 1982. "Super mice" were genetically altered to produce the human drug, tPA (tissue plasminogen activator to treat blood clots), in 1987. Since then, "super mice" pharming has come a long way. Using RNA interference, scientists have produced a cow whose milk contains increased amounts of casein, a protein used to make cheese and other foods, and almost no beta-lactoglobulin, a component in milk whey protein that causes allergies.

Pharming examples:

- Haemoglobin as a blood substitute
- Human protein C anticoagulant
- Alpha-1 antitrypsin (AAT) for treatment of AAT deficiency
- Insulin for diabetes treatment
- Vaccines (antigens)
- Growth hormones for treatment of deficiencies
- Factor VIII blood clotting factor

- Factor IX blood clotting factor

- Fibrinogen blood clotting factor

- Lactoferrin as an infant formula additive

Medical

Transgenesis can be used to neutralize genes that would normally prevent xenotransplantation. For example, a protein found in pigs can cause humans to reject their transplanted organs. This protein can be replaced by a similar human genome to prevent the rejection.

Ethical Concerns

Transgenesis has created certain ethical concerns. Examples include rights for animals that have been improved intellectually, legal ramifications, and possible health risks.

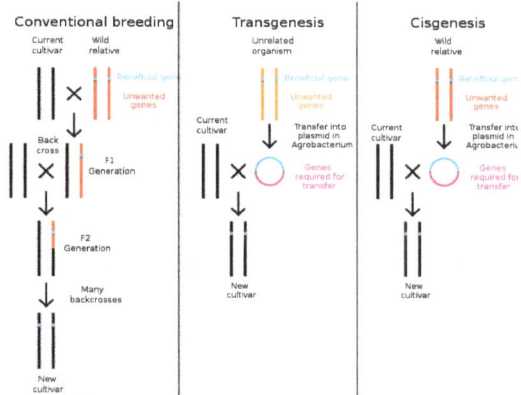

A diagram comparing the genetic changes achieved through conventional plant breeding, transgenesis and cisgenesis

New genotypes created with transgenic technologies also require multiple backcrossings. Furthermore, backcrossing does not account for the majority of time required to create, field test and release/commercialize a new variety.

Horizontal Gene Transfer

Horizontal gene transfer (HGT) or lateral gene transfer (LGT) is the movement of genetic material between unicellular and/or multicellular organisms other than by the ("vertical") transmission of DNA from parent to offspring. HGT is an important factor in the evolution of many organisms.

Horizontal gene transfer is the primary mechanism for the spread of antibiotic resistance in bacteria, plays an important role in the evolution of bacteria that can degrade novel compounds such as human-created pesticides and in the evolution, maintenance, and transmission of virulence.

It often involves temperate bacteriophages and plasmids. Genes responsible for antibiotic resistance in one species of bacteria can be transferred to another species of bacteria through various mechanisms such as F-pilus, subsequently arming the antibiotic resistant genes' recipient against antibiotics, which is becoming a medical challenge to deal with.

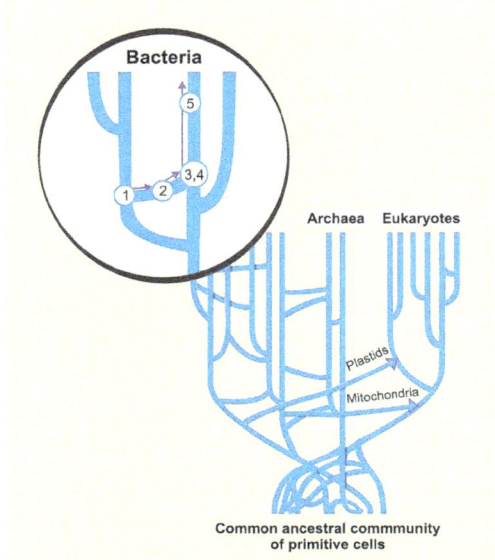

Tree of life showing vertical and horizontal gene transfers

Most thinking in genetics has focused upon vertical transfer, but horizontal gene transfer is important, and among single-celled organisms is perhaps the dominant form of genetic transfer.

Artificial horizontal gene transfer is a form of genetic engineering.

History

Horizontal genetic transfer was first described in Seattle in 1951, in a paper demonstrating that the transfer of a viral gene into *Corynebacterium diphtheriae* created a virulent strain from a non-virulent strain, also simultaneously solving the riddle of diphtheria (that patients could be infected with the bacteria but not have any symptoms, and then suddenly convert later or never), and giving the first example for the relevance of the lysogenic cycle. Inter-bacterial gene transfer was first described in Japan in a 1959 publication that demonstrated the transfer of antibiotic resistance between different species of bacteria. In the mid-1980s, Syvanen predicted that lateral gene transfer existed, had biological significance, and was involved in shaping evolutionary history from the beginning of life on Earth.

As Jian, Rivera and Lake (1999) put it: "Increasingly, studies of genes and genomes are indicating that considerable horizontal transfer has occurred between prokaryotes". The phenomenon appears to have had some significance for unicellular eukaryotes as well. As Bapteste et al. (2005) observe, "additional evidence suggests that gene transfer might also be an important evolutionary mechanism in protist evolution."

There is some evidence that even higher plants and animals have been affected and this has raised concerns for safety. Grafting of one plant to another can transfer chloroplasts (organelles in plant cells that conduct photosynthesis), mitochondrial DNA, and the entire cell nucleus containing the genome to potentially make a new species. Some Lepidoptera (e.g. monarch butterflies and silk-worms) have been genetically modified by horizontal gene transfer from the wasp bracovirus. Bites from the insect Reduviidae (assassin bug) can, via a parasite, infect humans with the trypanosomal Chagas disease, which can insert its DNA into the human genome. It has been suggested that lateral gene transfer to humans from bacteria may play a role in cancer.

Richardson and Palmer (2007) state: "Horizontal gene transfer (HGT) has played a major role in bacterial evolution and is fairly common in certain unicellular eukaryotes. However, the prevalence and importance of HGT in the evolution of multicellular eukaryotes remain unclear."

Due to the increasing amount of evidence suggesting the importance of these phenomena for evolution molecular biologists such as Peter Gogarten have described horizontal gene transfer as "A New Paradigm for Biology".

Some have argued that the process may be a hidden hazard of genetic engineering as it could allow transgenic DNA to spread from species to species.

Mechanism

There are several mechanisms for horizontal gene transfer:

- Transformation, the genetic alteration of a cell resulting from the introduction, uptake and expression of foreign genetic material (DNA or RNA). This process is relatively common in bacteria, but less so in eukaryotes. Transformation is often used in laboratories to insert novel genes into bacteria for experiments or for industrial or medical applications.

- Transduction, the process in which bacterial DNA is moved from one bacterium to another by a virus (a bacteriophage, or phage).

- Bacterial conjugation, a process that involves the transfer of DNA via a plasmid from a donor cell to a recombinant recipient cell during cell-to-cell contact.

- Gene transfer agents, virus-like elements encoded by the host that are found in the alphaproteobacteria order Rhodobacterales.

A transposon (jumping gene) is a mobile segment of DNA that can sometimes pick up a resistance gene and insert it into a plasmid or chromosome, thereby inducing horizontal gene transfer of antibiotic resistance.

Inference

Horizontal gene transfer is typically inferred using bioinformatic methods, either by identifying atypical sequence signatures ("parametric" methods) or by identifying strong discrepancies between the evolutionary history of particular sequences compared to that of their hosts.

Viruses

The virus called *Mimivirus* infects amoebae. Another virus, called *Sputnik*, also infects amoebae, but it cannot reproduce unless mimivirus has already infected the same cell. "Sputnik's genome reveals further insight into its biology. Although 13 of its genes show little similarity to any other known genes, three are closely related to mimivirus and mamavirus genes, perhaps cannibalized by the tiny virus as it packaged up particles sometime in its history. This suggests that the satellite virus could perform horizontal gene transfer between viruses, paralleling the way that bacterio-phages ferry genes between bacteria." Horizontal transfer is also seen between geminiviruses and tobacco plants.

Prokaryotes

Horizontal gene transfer is common among bacteria, even among very distantly related ones. This process is thought to be a significant cause of increased drug resistance when one bacterial cell acquires resistance, and the resistance genes are transferred to other species. Transposition and horizontal gene transfer, along with strong natural selective forces have led to multi-drug resistant strains of *S. aureus* and many other pathogenic bacteria. Horizontal gene transfer also plays a role in the spread of virulence factors, such as exotoxins and exoenzymes, amongst bacteria. A prime example concerning the spread of exotoxins is the adaptive evolution of Shiga toxins in *E. coli* through horizontal gene transfer via transduction with *Shigella* species of bacteria. Strategies to combat certain bacterial infections by targeting these specific virulence factors and mobile genetic elements have been proposed. For example, horizontally transferred genetic elements play important roles in the virulence of *E. coli*, *Salmonella*, *Streptococcus* and *Clostridium perfringens*.

In prokaryotes, restriction-modification systems are known to provide immunity against horizontal gene transfer and in stabilizing mobile genetic elements. Genes encoding restriction modification systems have been reported to move between prokaryotic genomes within mobile genetic elements such as plasmids, prophages, insertion sequences/transposons, integrative conjugative elements (ICEs), and integrons. Still, they are more frequently a chromosomal-encoded barrier to MGEs than an MGE-encoded tool for cell infection.

Bacterial Transformation

Natural transformation is a bacterial adaptation for DNA transfer (HGT) that depends on the expression of numerous bacterial genes whose products are responsible for this process. In general, transformation is a complex, energy-requiring developmental process. In order for a bacterium to bind, take up and recombine exogenous DNA into its chromosome, it must become competent, that is, enter a special physiological state. Competence development in *Bacillus subtilis* requires expression of about 40 genes. The DNA integrated into the host chromosome is usually (but with infrequent exceptions) derived from another bacterium of the same species, and is thus homologous to the resident chromosome. The capacity for natural transformation occurs in at least 67 prokaryotic species. Competence for transformation is typically induced by high cell density and/or nutritional limitation, conditions associated with the stationary phase of bacterial growth. Competence appears to be an adaptation for DNA repair. Transformation in bacteria can be viewed as a primitive sexual process, since it involves interaction of homologous DNA from two individuals to form recombinant DNA that is passed on to succeeding generations. Although transduction is the form of HGT most commonly associated with bacteriophages, certain phages may also be able to promote transformation.

Bacterial Conjugation

Conjugation in *Mycobacterium smegmatis*, like conjugation in *E. coli*, requires stable and extended contact between a donor and a recipient strain, is DNase resistant, and the transferred DNA is incorporated into the recipient chromosome by homologous recombination. However, unlike *E. coli* high frequency of recombination conjugation (Hfr), mycobacterial conjugation is a type of HGT that is chromosome rather than plasmid based. Furthermore, in contrast to *E. coli* (Hfr) conjugation, in *M. smegmatis* all regions of the chromosome are transferred with comparable efficiencies. Substantial blending of the parental genomes was found as a result of conjugation, and this blending was regarded as reminiscent of that seen in the meiotic products of sexual reproduction.

Archaeal DNA Transfer

The archaeon *Sulfolobus solfataricus*, when UV irradiated, strongly induces the formation of type IV pili which then facilitates cellular aggregation. Exposure to chemical agents that cause DNA damage also induces cellular aggregation. Other physical stressors, such as temperature shift or pH, do not induce aggregation, suggesting that DNA damage is a specific inducer of cellular aggregation.

UV-induced cellular aggregation mediates intercellular chromosomal HGT marker exchange with high frequency, and UV-induced cultures display recombination rates that exceed those of uninduced cultures by as much as three orders of magnitude. *S. solfataricus* cells aggregate preferentially with other cells of their own species. Frols et al. and Ajon et al. suggested that UV-inducible DNA transfer is likely an important mechanism for providing increased repair of damaged DNA via homologous recombination. This process can be regarded as a simple form of sexual interaction.

Another thermophilic species, *Sulfolobus acidocaldarius*, is able to undergo HGT. *S. acidocaldarius* can exchange and recombine chromosomal markers at temperatures up to 84°C. UV exposure induces pili formation and cellular aggregation. Cells with the ability to aggregate have greater survival than mutants lacking pili that are unable to aggregate. The frequency of recombination is increased by DNA damage induced by UV-irradiation and by DNA damaging chemicals.

The *ups* operon, containing five genes, is highly induced by UV irradiation. The proteins encoded by the *ups* operon are employed in UV-induced pili assembly and cellular aggregation leading to intercellular DNA exchange and homologous recombination. Since this system increases the fitness of *S. acidocaldarius* cells after UV exposure, Wolferen et al. considered that transfer of DNA likely takes place in order to repair UV-induced DNA damages by homologous recombination.

Eukaryotes

"Sequence comparisons suggest recent horizontal transfer of many genes among diverse species including across the boundaries of phylogenetic 'domains'. Thus determining the phylogenetic history of a species can not be done conclusively by determining evolutionary trees for single genes."

- Analysis of DNA sequences suggests that horizontal gene transfer has occurred within eukaryotes from the chloroplast and mitochondrial genomes to the nuclear genome. As stated in the endosymbiotic theory, chloroplasts and mitochondria probably originated as bacterial endosymbionts of a progenitor to the eukaryotic cell.

- Horizontal transfer occurs from bacteria to some fungi, such as the yeast *Saccharomyces cerevisiae*.

- The adzuki bean beetle has acquired genetic material from its (non-beneficial) endosymbiont *Wolbachia*. New examples have recently been reported demonstrating that Wolbachia bacteria represent an important potential source of genetic material in arthropods and filarial nematodes.

- Mitochondrial genes moved to parasites of the Rafflesiaceae plant family from their hosts and from chloroplasts of a still-unidentified plant to the mitochondria of the bean *Phaseolus*.

- *Striga hermonthica*, a parasitic eudicot, has received a gene from sorghum (*Sorghum bicolor*) to its nuclear genome. The gene's functionality is unknown.

- Pea aphids (*Acyrthosiphon pisum*) contain multiple genes from fungi. Plants, fungi, and microorganisms can synthesize carotenoids, but torulene made by pea aphids is the only carotenoid known to be synthesized by an organism in the animal kingdom.

- The malaria pathogen *Plasmodium vivax* acquired genetic material from humans that might help facilitate its long stay in the body.

- A bacteriophage-mediated mechanism transfers genes between prokaryotes and eukaryotes. Nuclear localization signals in bacteriophage terminal proteins (TP) prime DNA replication and become covalently linked to the viral genome. The role of virus and bacteriophages in HGT in bacteria, suggests that TP-containing genomes could be a vehicle of inter-kingdom genetic information transference all throughout evolution.

- HhMAN1 is a gene in the genome of the coffee borer beetle (*Hypothenemus hampei*) that resembles bacterial genes, and is thought to be transferred from bacteria in the beetle's gut.

- A gene that allowed ferns to survive in dark forests came from the hornwort, which grows in mats on streambanks or trees. The neochrome gene arrived about 180 million years ago.

- Plants are capable of receiving genetic information from viruses by horizontal gene transfer.

- One study identified approximately 100 of humans' approximately 20,000 total genes which likely resulted from horizontal gene transfer, but this number has been challenged by several researchers arguing these candidate genes for HGT are more likely the result of gene loss combined with differences in the rate of evolution

- Bdelloid rotifers currently hold the 'record' for HGT in animals with ~8% of their genes from bacterial origins. Tardigrades were thought to break the record with 17.5% HGT, but that finding was an artifact of bacterial contamination.

- A study found the genomes of 40 animals (including 10 primates, four *Caenorhabditis* worms, and 12 *Drosophila* insects) contained genes which the researchers concluded had been transferred from bacteria and fungi by horizontal gene transfer. The researchers estimated that for some nematodes and Drosophilia insects these genes had been acquired relatively recently.

- The eastern emerald sea slug *Elysia chlorotica* has been suggested by FISH analysis to contain photosynthesis-supporting genes obtained from an algae (*Vaucheria litorea*) in their diet.

Horizontal Transposon Transfer

Horizontal transposon transfer (HTT) refers to the passage of pieces of DNA that are characterized by their ability to move from one locus to another between genomes by means other than parent-to-offspring inheritance. Horizontal gene transfer has long been thought to be crucial to prokaryotic evolution, but there is a growing amount of data showing that HTT is a common and widespread phenomenon in eukaryote evolution as well. On the transposable element (TE) side, spreading between genomes via horizontal transfer may be viewed as a strategy to escape purging due to purifying selection, mutational decay and/or host defense mechanisms.

HTT can occur with any type of transposable elements, but DNA transposons and LTR retroelements are more likely to be capable of HTT because both have a stable, double-stranded DNA intermediate that is thought to be sturdier than the single-stranded RNA intermediate of non-LTR retroelements, which can be highly degradable. Non-autonomous elements may be less likely to transfer horizontally compared to autonomous elements because they do not encode the proteins required for their own mobilization. The structure of these non-autonomous elements generally consists of an intronless gene encoding a transposase protein, and may or may not have a promoter sequence. Those that do not have promoter sequences encoded within the mobile region rely on adjacent host promoters for expression. Horizontal transfer is thought to play an important role in the TE life cycle.

HTT has been shown to occur between species and across continents in both plants and animals (Ivancevic et al. 2013), though some TEs have been shown to more successfully colonize the genomes of certain species over others. Both spatial and taxonomic proximity of species has been proposed to favor HTTs in plants and animals. It is unknown how the density of a population may affect the rate of HTT events within a population, but close proximity due to parasitism and cross contamination due to crowding have been proposed to favor HTT in both plants and animals. Successful transfer of a transposable element requires delivery of DNA from donor to host cell (and to the germ line for multi-cellular organisms), followed by integration into the recipient host genome. Though the actual mechanism for the transportation of TEs from donor cells to host cells is unknown, it is established that naked DNA and RNA can circulate in bodily fluid. Many proposed vectors include arthropods, viruses, freshwater snails (Ivancevic et al. 2013), endosymbiotic bacteria, and intracellular parasitic bacteria. In some cases, even TEs facilitate transport for other TEs.

The arrival of a new TE in a host genome can have detrimental consequences because TE mobility may induce mutation. However, HTT can also be beneficial by introducing new genetic material into a genome and promoting the shuffling of genes and TE domains among hosts, which can be co-opted by the host genome to perform new functions. Moreover, transposition activity increases the TE copy number and generates chromosomal rearrangement hotspots. HTT detection is a difficult task because it is an ongoing phenomenon that is constantly changing in frequency of occurrence and composition of TEs inside host genomes. Furthermore, few species have been analyzed for HTT, making it difficult to establish patterns of HTT events between species. These issues can lead to the underestimation or overestimation of HTT events between ancestral and current eukaryotic species.

Artificial Horizontal Gene Transfer

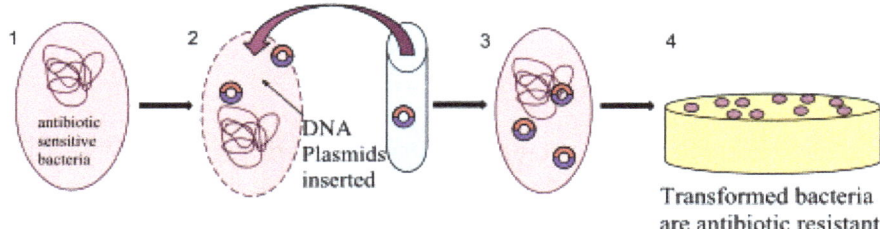

Before it is transformed a bacterium is susceptible to antibiotics. A plasmid can be inserted when the bacteria is under stress, and be incorporated into the bacterial DNA creating antibiotic resistance. When the plasmids are prepared they are inserted into the bacterial cell by either making pores in the plasma membrane with temperature extremes and chemical treatments, or making it semi permeable through the process of electrophoresis, in which electric currents create the holes in the membrane. After conditions return to normal the holes in the membrane close and the plasmids are trapped inside the bacteria where they become part of the genetic material and their genes are expressed by the bacteria.

Genetic engineering is essentially horizontal gene transfer, albeit with synthetic expression cassettes. The Sleeping Beauty transposon system (SB) was developed as a synthetic gene transfer agent that was based on the known abilities of Tc1/mariner transposons to invade genomes of extremely diverse species. The SB system has been used to introduce genetic sequences into a wide variety of animal genomes.

Importance in Evolution

Horizontal gene transfer is a potential confounding factor in inferring phylogenetic trees based on the sequence of one gene. For example, given two distantly related bacteria that have exchanged a gene a phylogenetic tree including those species will show them to be closely related because that gene is the same even though most other genes are dissimilar. For this reason it is often ideal to use other information to infer robust phylogenies such as the presence or absence of genes or, more commonly, to include as wide a range of genes for phylogenetic analysis as possible.

For example, the most common gene to be used for constructing phylogenetic relationships in prokaryotes is the 16S ribosomal RNA gene since its sequences tend to be conserved among members with close phylogenetic distances, but variable enough that differences can be measured. However, in recent years it has also been argued that 16s rRNA genes can also be horizontally transferred. Although this may be infrequent, the validity of 16s rRNA-constructed phylogenetic trees must be reevaluated.

Biologist Johann Peter Gogarten suggests "the original metaphor of a tree no longer fits the data from recent genome research" therefore "biologists should use the metaphor of a mosaic to describe the different histories combined in individual genomes and use the metaphor of a net to visualize the rich exchange and cooperative effects of HGT among microbes". There exist several methods to infer such phylogenetic networks.

Using single genes as phylogenetic markers, it is difficult to trace organismal phylogeny in the presence of horizontal gene transfer. Combining the simple coalescence model of cladogenesis with rare HGT horizontal gene transfer events suggest there was no single most recent common ancestor that contained all of the genes ancestral to those shared among the three domains of life. Each contemporary molecule has its own history and traces back to an individual molecule cenancestor. However, these molecular ancestors were likely to be present in different organisms at different times."

Challenge to The Tree Of Life

Horizontal gene transfer poses a possible challenge to the concept of the last universal common ancestor (LUCA) at the root of the tree of life first formulated by Carl Woese, which led him to propose the Archaea as a third domain of life. Indeed, it was while examining the new three-domain view of life that horizontal gene transfer arose as a complicating issue: *Archaeoglobus fulgidus* was seen as an anomaly with respect to a phylogenetic tree based upon the encoding for the enzyme HMGCoA reductase—the organism in question is a definite Archaean, with all the cell lipids and transcription machinery that are expected of an Archaean, but whose HMGCoA genes are of bacterial origin. Scientists are broadly agreed on symbiogenesis, that mitochondria in eukaryotes derived from alpha-proteobacterial cells and that chloroplasts came from ingested cyanobacteria, and other gene transfers may have affected early eukaryotes. (In contrast, multicellular eukaryotes have mechanisms to prevent horizontal gene transfer, including separated germ cells.) If there had been continued and extensive gene transfer, there would be a complex network with many ancestors, instead of a tree of life with sharply delineated lineages leading back to a LUCA. However, a LUCA can be identified, so horizontal transfers must have been relatively limited.

Genes

There is evidence for historical horizontal transfer of the following genes:

- Lycopene cyclase for carotenoid biosynthesis, between Chlorobi and Cyanobacteria.

- *TetO* gen conferring resistance to tetracycline, between *Campylobacter jejuni*.

- Neochrome, gene in some ferns that enhances their ability to survive in dim light. Believed to have been acquired from algae sometime during the Cretaceous.

- transfer of a cysteine synthase from a bacterium into phytophagous mites and Lepidoptera allowing the detoxification of cyanogenic glucosides produced by host plants.

- The LINE1 sequence has transferred from humans to the gonorrhea bacteria.

Molecular Cloning

Molecular cloning is a set of experimental methods in molecular biology that are used to assemble recombinant DNA molecules and to direct their replication within host organisms. The use of the word *cloning* refers to the fact that the method involves the replication of one molecule to produce a population of cells with identical DNA molecules. Molecular cloning generally uses DNA sequences from two different organisms: the species that is the source of the DNA to be cloned, and the species that will serve as the living host for replication of the recombinant DNA. Molecular cloning methods are central to many contemporary areas of modern biology and medicine.

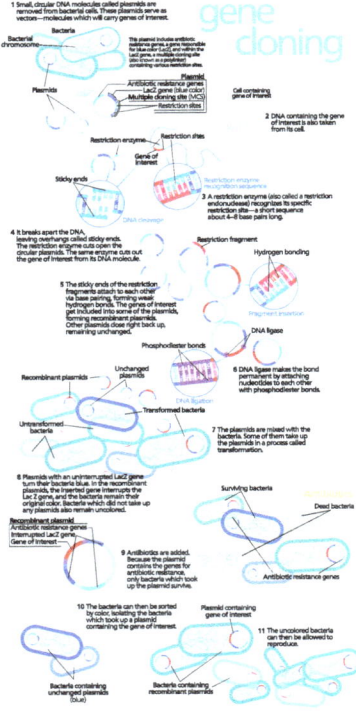

Diagram of molecular cloning using bacteria and plasmids.

In a conventional molecular cloning experiment, the DNA to be cloned is obtained from an organism of interest, then treated with enzymes in the test tube to generate smaller DNA fragments. Subsequently, these fragments are then combined with vector DNA to generate recombinant DNA molecules. The recombinant DNA is then introduced into a host organism (typically an easy-to-grow, benign, laboratory strain of *E. coli* bacteria). This will generate a population of organisms in which recombinant DNA molecules are replicated along with the host DNA. Because they contain foreign DNA fragments, these are transgenic or genetically modified microorganisms (GMO). This process takes advantage of the fact that a single bacterial cell can be induced to take up and replicate a single recombinant DNA molecule. This single cell can then be expanded exponentially to generate a large amount of bacteria, each of which contain copies of the original recombinant molecule. Thus, both the resulting bacterial population, and the recombinant DNA molecule, are commonly referred to as "clones". Strictly speaking, *recombinant DNA* refers to DNA molecules, while *molecular cloning* refers to the experimental methods used to assemble them. The idea arose that different DNA sequences could be inserted into a plasmid and that these foreign sequences would be carried into bacteria and digest as part of the plasmid. That is, these plasmids could serve as cloning vectors to carry genes. It could insert different base pair and cause other mutation in bacterial gene.

Virtually any DNA sequence can be cloned and amplified, but there are some factors that might limit the success of the process. Examples of the DNA sequences that are difficult to clone are inverted repeats, origins of replication, centimeters and telomeres. Another characteristic that limits chances of success is large size of DNA sequence. Inserts larger than 10kbp have very limited success, but bacteriophages such as bacteriophage λ can be modified to successfully insert a sequence up to 40 kbp.

History of Molecular Cloning

Prior to the 1970s, our understanding of genetics and molecular biology was severely hampered by an inability to isolate and study individual genes from complex organisms. This changed dramatically with the advent of molecular cloning methods. Microbiologists, seeking to understand the molecular mechanisms through which bacteria restricted the growth of bacteriophage, isolated restriction endonucleases, enzymes that could cleave DNA molecules only when specific DNA sequences were encountered. They showed that restriction enzymes cleaved chromosome-length DNA molecules at specific locations, and that specific sections of the larger molecule could be purified by size fractionation. Using a second enzyme, DNA ligase, fragments generated by restriction enzymes could be joined in new combinations, termed recombinant DNA. By recombining DNA segments of interest with vector DNA, such as bacteriophage or plasmids, which naturally replicate inside bacteria, large quantities of purified recombinant DNA molecules could be produced in bacterial cultures. The first recombinant DNA molecules were generated and studied in 1972.

Overview

Molecular cloning takes advantage of the fact that the chemical structure of DNA is fundamentally the same in all living organisms. Therefore, if any segment of DNA from any organism is inserted into a DNA segment containing the molecular sequences required for DNA replication, and the resulting recombinant DNA is introduced into the organism from which the replication sequences were obtained, then the foreign DNA will be replicated along with the host cell's DNA in the transgenic organism.

Molecular cloning is similar to polymerase chain reaction (PCR) in that it permits the replication of DNA sequence. The fundamental difference between the two methods is that molecular cloning involves replication of the DNA in a living microorganism, while PCR replicates DNA in an *in vitro* solution, free of living cells.

Steps in Molecular Cloning

In standard molecular cloning experiments, the cloning of any DNA fragment essentially involves seven steps: (1) Choice of host organism and cloning vector, (2) Preparation of vector DNA, (3) Preparation of DNA to be cloned, (4) Creation of recombinant DNA, (5) Introduction of recombinant DNA into host organism, (6) Selection of organisms containing recombinant DNA, (7) Screening for clones with desired DNA inserts and biological properties.

Although the detailed planning of the cloning can be done in any text editor, together with online utilities for e.g. PCR primer design, dedicated software exist for the purpose. Software for the purpose include for example ApE (open source), DNAStrider (open source), Serial Cloner (gratis) and Collagene (open source).

Notably, the growing capacity and fidelity of DNA synthesis platforms allows for increasingly intricate designs in molecular engineering. These projects may include very long strands of novel DNA sequence and/or test entire libraries simultaneously, as opposed to of individual sequences. These shifts introduce complexity require design to move away from the flat nucleotide based representation and towards a higher level of abstraction. Examples of such tools are GenoCAD GenoCAD, Teselagen (free for academia) or GeneticConstructor (free for academics).

Choice of Host Organism and Cloning Vector

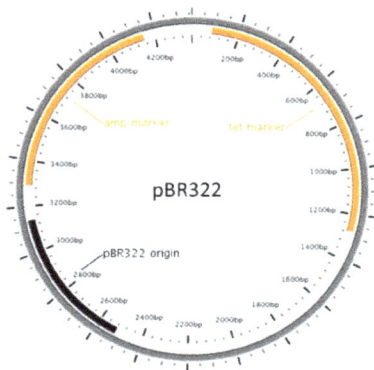

Diagram of a commonly used cloning plasmid; pBR322. It's a circular piece of DNA 4361 bases long. Two antibiotic resistance genes are present, conferring resistance to ampicillin and tetracycline, and an origin of replication that the host uses to replicate the DNA.

Although a very large number of host organisms and molecular cloning vectors are in use, the great majority of molecular cloning experiments begin with a laboratory strain of the bacterium *E. coli* (*Escherichia coli*) and a plasmid cloning vector. *E. coli* and plasmid vectors are in common use because they are technically sophisticated, versatile, widely available, and offer rapid growth of recombinant organisms with minimal equipment. If the DNA to be cloned is exceptionally large (hundreds of thousands to millions of base pairs), then a bacterial artificial chromosome or yeast artificial chromosome vector is often chosen.

Specialized applications may call for specialized host-vector systems. For example, if the experimentalists wish to harvest a particular protein from the recombinant organism, then an expression vector is chosen that contains appropriate signals for transcription and translation in the desired host organism. Alternatively, if replication of the DNA in different species is desired (for example, transfer of DNA from bacteria to plants), then a multiple host range vector (also termed shuttle vector) may be selected. In practice, however, specialized molecular cloning experiments usually begin with cloning into a bacterial plasmid, followed by subcloning into a specialized vector.

Whatever combination of host and vector are used, the vector almost always contains four DNA segments that are critically important to its function and experimental utility:

1. DNA *replication origin* is necessary for the vector (and its linked recombinant sequences) to replicate inside the host organism.

2. one or more unique *restriction endonuclease recognition sites* to serves as sites where foreign DNA may be introduced.

3. a *selectable genetic marker* gene that can be used to enable the survival of cells that have taken up vector sequences.

4. a *tag* gene that can be used to screen for cells containing the foreign DNA.

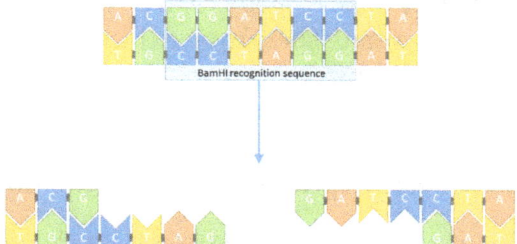

Cleavage of a DNA sequence containing the BamHI restriction site. The DNA is cleaved at the palindromic sequence to produce 'sticky ends'.

Preparation of Vector DNA

The cloning vector is treated with a restriction endonuclease to cleave the DNA at the site where foreign DNA will be inserted. The restriction enzyme is chosen to generate a configuration at the cleavage site that is compatible with the ends of the foreign DNA. Typically, this is done by cleaving the vector DNA and foreign DNA with the same restriction enzyme, for example EcoRI. Most modern vectors contain a variety of convenient cleavage sites that are unique within the vector molecule (so that the vector can only be cleaved at a single site) and are located within a gene (frequently beta-galactosidase) whose inactivation can be used to distinguish recombinant from non-recombinant organisms at a later step in the process. To improve the ratio of recombinant to non-recombinant organisms, the cleaved vector may be treated with an enzyme (alkaline phosphatase) that dephosphorylates the vector ends. Vector molecules with dephosphorylated ends are unable to replicate, and replication can only be restored if foreign DNA is integrated into the cleavage site.

Preparation of DNA to be Cloned

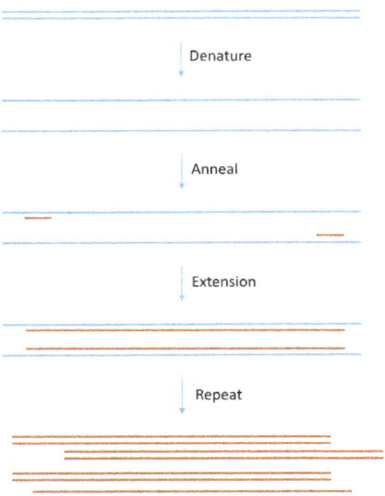

DNA for cloning is most commonly produced using PCR. Template DNA is mixed with bases (the building blocks of DNA), primers (short pieces of complementary single stranded DNA) and a DNA polymerase enzyme that builds the DNA chain. The mix goes through cycles of heating and cooling to produce large quantities of copied DNA.

For cloning of genomic DNA, the DNA to be cloned is extracted from the organism of interest. Virtually any tissue source can be used (even tissues from extinct animals), as long as the DNA is not extensively degraded. The DNA is then purified using simple methods to remove contaminating proteins (extraction with phenol), RNA (ribonuclease) and smaller molecules (precipitation and/or chromatography). Polymerase chain reaction (PCR) methods are often used for amplification of specific DNA or RNA (RT-PCR) sequences prior to molecular cloning.

DNA for cloning experiments may also be obtained from RNA using reverse transcriptase (complementary DNA or cDNA cloning), or in the form of synthetic DNA (artificial gene synthesis). cDNA cloning is usually used to obtain clones representative of the mRNA population of the cells of interest, while synthetic DNA is used to obtain any precise sequence defined by the designer.

The purified DNA is then treated with a restriction enzyme to generate fragments with ends capable of being linked to those of the vector. If necessary, short double-stranded segments of DNA (*linkers*) containing desired restriction sites may be added to create end structures that are compatible with the vector.

Creation of Recombinant DNA with DNA Ligase

The creation of recombinant DNA is in many ways the simplest step of the molecular cloning process. DNA prepared from the vector and foreign source are simply mixed together at appropriate concentrations and exposed to an enzyme (DNA ligase) that covalently links the ends together. This joining reaction is often termed ligation. The resulting DNA mixture containing randomly joined ends is then ready for introduction into the host organism.

DNA ligase only recognizes and acts on the ends of linear DNA molecules, usually resulting in a complex mixture of DNA molecules with randomly joined ends. The desired products (vector DNA covalently linked to foreign DNA) will be present, but other sequences (e.g. foreign DNA linked to itself, vector DNA linked to itself and higher-order combinations of vector and foreign DNA) are also usually present. This complex mixture is sorted out in subsequent steps of the cloning process, after the DNA mixture is introduced into cells.

Introduction of Recombinant DNA into Host Organism

The DNA mixture, previously manipulated in vitro, is moved back into a living cell, referred to as the host organism. The methods used to get DNA into cells are varied, and the name applied to this step in the molecular cloning process will often depend upon the experimental method that is chosen (e.g. transformation, transduction, transfection, electroporation).

When microorganisms are able to take up and replicate DNA from their local environment, the process is termed transformation, and cells that are in a physiological state such that they can take up DNA are said to be competent. In mammalian cell culture, the analogous process of introducing DNA into cells is commonly termed transfection. Both transformation and transfection usually require preparation of the cells through a special growth regime and chemical treatment process that will vary with the specific species and cell types that are used.

Electroporation uses high voltage electrical pulses to translocate DNA across the cell membrane (and cell wall, if present). In contrast, transduction involves the packaging of DNA into virus-derived particles, and using these virus-like particles to introduce the encapsulated DNA into the cell through a process resembling viral infection. Although electroporation and transduction are highly specialized methods, they may be the most efficient methods to move DNA into cells.

Selection of Organisms Containing Vector Sequences

Whichever method is used, the introduction of recombinant DNA into the chosen host organism is usually a low efficiency process; that is, only a small fraction of the cells will actually take up DNA. Experimental scientists deal with this issue through a step of artificial genetic selection, in which cells that have not taken up DNA are selectively killed, and only those cells that can actively replicate DNA containing the selectable marker gene encoded by the vector are able to survive.

When bacterial cells are used as host organisms, the selectable marker is usually a gene that confers resistance to an antibiotic that would otherwise kill the cells, typically ampicillin. Cells harboring the plasmid will survive when exposed to the antibiotic, while those that have failed to take up plasmid sequences will die. When mammalian cells (e.g. human or mouse cells) are used, a similar strategy is used, except that the marker gene (in this case typically encoded as part of the kanMX cassette) confers resistance to the antibiotic Geneticin.

Screening for Clones with Desired DNA Inserts and Biological Properties

Modern bacterial cloning vectors (e.g. pUC19 and later derivatives including the pGEM vectors) use the blue-white screening system to distinguish colonies (clones) of transgenic cells from those that contain the parental vector (i.e. vector DNA with no recombinant sequence inserted). In these vectors, foreign DNA is inserted into a sequence that encodes an essential part of beta-galactosidase, an enzyme whose activity results in formation of a blue-colored colony on the culture medium that is used for this work. Insertion of the foreign DNA into the beta-galactosidase coding sequence disables the function of the enzyme, so that colonies containing transformed DNA remain colorless (white). Therefore, experimentalists are easily able to identify and conduct further studies on transgenic bacterial clones, while ignoring those that do not contain recombinant DNA.

The total population of individual clones obtained in a molecular cloning experiment is often termed a DNA library. Libraries may be highly complex (as when cloning complete genomic DNA from an organism) or relatively simple (as when moving a previously cloned DNA fragment into a different plasmid), but it is almost always necessary to examine a number of different clones to be sure that the desired DNA construct is obtained. This may be accomplished through a very wide range of experimental methods, including the use of nucleic acid hybridizations, antibody probes, polymerase chain reaction, restriction fragment analysis and/or DNA sequencing.

Applications of Molecular Cloning

Molecular cloning provides scientists with an essentially unlimited quantity of any individual DNA segments derived from any genome. This material can be used for a wide range of purposes, including those in both basic and applied biological science. A few of the more important applications are summarized here.

Genome Organization and Gene Expression

Molecular cloning has led directly to the elucidation of the complete DNA sequence of the genomes of a very large number of species and to an exploration of genetic diversity within individual species, work that has been done mostly by determining the DNA sequence of large numbers of randomly cloned fragments of the genome, and assembling the overlapping sequences.

At the level of individual genes, molecular clones are used to generate probes that are used for examining how genes are expressed, and how that expression is related to other processes in biology, including the metabolic environment, extracellular signals, development, learning, senescence and cell death. Cloned genes can also provide tools to examine the biological function and importance of individual genes, by allowing investigators to inactivate the genes, or make more subtle mutations using regional mutagenesis or site-directed mutagenesis.

Production of Recombinant Proteins

Obtaining the molecular clone of a gene can lead to the development of organisms that produce the protein product of the cloned genes, termed a recombinant protein. In practice, it is frequently more difficult to develop an organism that produces an active form of the recombinant protein in desirable quantities than it is to clone the gene. This is because the molecular signals for gene expression are complex and variable, and because protein folding, stability and transport can be very challenging.

Many useful proteins are currently available as recombinant products. These include--(1) medically useful proteins whose administration can correct a defective or poorly expressed gene (e.g. recombinant factor VIII, a blood-clotting factor deficient in some forms of hemophilia, and recombinant insulin, used to treat some forms of diabetes), (2) proteins that can be administered to assist in a life-threatening emergency (e.g. tissue plasminogen activator, used to treat strokes), (3) recombinant subunit vaccines, in which a purified protein can be used to immunize patients against infectious diseases, without exposing them to the infectious agent itself (e.g. hepatitis B vaccine), and (4) recombinant proteins as standard material for diagnostic laboratory tests.

Transgenic Organisms

Once characterized and manipulated to provide signals for appropriate expression, cloned genes may be inserted into organisms, generating transgenic organisms, also termed genetically modified organisms (GMOs). Although most GMOs are generated for purposes of basic biological research, a number of GMOs have been developed for commercial use, ranging from animals and plants that produce pharmaceuticals or other compounds (pharming), herbicide-resistant crop plants, and fluorescent tropical fish (GloFish) for home entertainment.

Gene Therapy

Gene therapy involves supplying a functional gene to cells lacking that function, with the aim of correcting a genetic disorder or acquired disease. Gene therapy can be broadly divided into two categories. The first is alteration of germ cells, that is, sperm or eggs, which results in a permanent genetic change for the whole organism and subsequent generations. This "germ line gene therapy" is considered by many to be unethical in human beings. The second type of gene therapy, "somatic cell gene therapy", is analogous to an organ transplant. In this case, one or more specific tissues are targeted by direct treatment or by removal of the tissue, addition of the therapeutic gene or genes in the laboratory, and return of the treated cells to the patient. Clinical trials of somatic cell gene therapy began in the late 1990s, mostly for the treatment of cancers and blood, liver, and lung disorders.

Despite a great deal of publicity and promises, the history of human gene therapy has been characterized by relatively limited success. The effect of introducing a gene into cells often promotes only partial and/or transient relief from the symptoms of the disease being treated. Some gene therapy trial patients have suffered adverse consequences of the treatment itself, including deaths. In some cases, the adverse effects result from disruption of essential genes within the patient's genome by insertional inactivation. In others, viral vectors used for gene therapy have been contaminated with infectious virus. Nevertheless, gene therapy is still held to be a promising future area of medicine, and is an area where there is a significant level of research and development activity.

References

- Schouten, H. J.; Jacobsen, E. (2007). "Are Mutations in Genetically Modified Plants Dangerous?". Journal of Biomedicine and Biotechnology. 2007: 1–2. doi:10.1155/2007/82612

- Bryan D. Ness, ed. (February 2004). "Transgenic Organisms". Encyclopedia of Genetics (Rev. ed.). Pacific Union College. ISBN 1-58765-149-1. Archived from the original on March 24, 2006

- Le Page, Michael (2016-03-17). "Farmers may have been accidentally making GMOs for millennia". The New Scientist. Retrieved 2016-07-11

- Mali P, Yang L, Esvelt KM, et al. (February 2013). "RNA-guided human genome engineering via Cas9". Science. 339 (6121): 823–6. PMC 3712628. PMID 23287722. doi:10.1126/science.1232033

- Mandel, Morton; Higa, Akiko (1970). "Calcium-dependent bacteriophage DNA infection". Journal of Molecular Biology. 53 (1): 159–162. PMID 4922220. doi:10.1016/0022-2836(70)90051-3

- Watson JD (2007). Recombinant DNA: genes and genomes: a short course. San Francisco: W.H. Freeman. ISBN 0-7167-2866-4

- Gruère, Colin A. Carter and Guillaume P. (2003-12-15). "Mandatory Labeling of Genetically Modified Foods: Does it Really Provide Consumer Choice?". www.agbioforum.org. Retrieved 2016-01-21

- Hwang WY, Fu Y, Reyon D, et al. (March 2013). "Efficient genome editing in zebrafish using a CRISPR-Cas system". Nat. Biotechnol. 31 (3): 227–9. PMC 3686313. PMID 23360964. doi:10.1038/nbt.2501

- Bratspies, Rebecca (2007). "Some Thoughts on the American Approach to Regulating Genetically Modified Organisms" (PDF). Kansas Journal of Law and Public Policy. 16: 393

- Patten CL, Glick BR, Pasternak J (2009). Molecular Biotechnology: Principles and Applications of Recombinant DNA. Washington, D.C: ASM Press. ISBN 978-1-55581-498-4

- Talbot, David (March 2016). "10 Breakthrough Technologies 2016: Precise Gene Editing in Plants". MIT Technology Review. Retrieved 2016-03-08

- Higuchi R, Bowman B, Freiberger M, Ryder OA, Wilson AC (1984). "DNA sequences from the quagga, an extinct member of the horse family". Nature. 312 (5991): 282–4. PMID 6504142. doi:10.1038/312282a0

- Gasdaska, John R.; Spencer, David; Dickey, Lynn (2003). "Advantages of Therapeutic Protein Production in the Aquatic Plant Lemna". BioProcessing Journal. 2 (2): 49–56. doi:10.12665/j22.gasdaska

- Russell DW, Sambrook J (2001). Molecular cloning: a laboratory manual. Cold Spring Harbor, N.Y: Cold Spring Harbor Laboratory. ISBN 978-0-87969-576-7

- "ProdiGene Launches First Large Scale-Up Manufacturing of Recombinant Protein From Plant System" (Press release). ProdiGene. February 13, 2002. Retrieved March 8, 2013

- Oldenburg J, Dolan G, Lemm G (Jan 2009). "Haemophilia care then, now and in the future". Haemophilia. 15 Suppl 1: 2–7. PMID 19125934. doi:10.1111/j.1365-2516.2008.01946.x

- Jagadeesan, Premananh (2015). "Transgenic and cloned animals in the food chain - are we prepared to tackle it?". Journal of the Science of Food and Agriculture. 95 (14): 2779–2782. PMID 25857482. doi:10.1002/jsfa.7205

- McHugen, Alan (14 September 2000). "Chapter 1: Hors-d'oeuvres and entrees/What is genetic modification? What are GMOs?". Pandora's Picnic Basket. Oxford University Press. ISBN 978-0198506744

- Hallenbeck, Terri (2014-04-27). "How GMO labeling came to pass in Vermont". Burlington Free Press. Retrieved 2014-05-28

- Lewandowski C, Barsan W (Feb 2001). "Treatment of acute ischemic stroke". Annals of Emergency Medicine. 37 (2): 202–16. PMID 11174240. doi:10.1067/mem.2001.111573

- EFSA (2010). "Guidance on the environmental risk assessment of genetically modified plants". EFSA Journal. 8 (11): 1879. doi:10.2903/j.efsa.2010.1879

- Mahgoub, Salah (2015). Genetically Modified Foods: Basics, Applications, and Controversy. Taylor & Francis Group. p. 9. ISBN 9781482242812

Genetic Engineering: Strategies, Methods and Techniques

Genetic engineering has several varying techniques which are used in modifying the genomes of an organism. The methods and techniques used in genetic engineering are gene knockin, gene targeting, gene knockout, electroporation and gene knockdown. This chapter discusses the methods and techniques of genetic engineering in a critical manner providing key analysis to the subject matter.

Genetic Engineering Techniques

Genetic engineering techniques enable modification of the DNA of living organisms. A variety of editing techniques have been developed since DNA's structure was first discovered.

Targets

Bacteria are commonly engineered for research purposes. Typically this is through transformation to add a plasmid containing a gene of interest, but editing of the chromosome is also used. Plants and animals have both been genetically modified for research, agricultural and medical applications. In plants, the most widely inserted genes provide herbicide resistance or insecticidal properties. In animals, the most widely used are growth hormone genes. Finally, genetically modified viruses (such as retroviruses and lentiviruses) are also used as viral vectors to transfer target genes to another organism in gene therapy.

Procedure

The first step involves choosing and isolating the gene that will be inserted into/removed from the genetically modified organism.

The gene must generally be combined with a promoter and terminator region as well as a selectable marker gene.

Then the genes must be spliced into the target's DNA. For animals, the gene must be inserted into embryonic stem cells.

The resulting organism must have the presence of the target gene confirmed.

First generation offspring are heterozygous, requiring them to be inbred to create the homozygous pattern necessary for stable inheritance. Homozygosity must be confirmed in second generation specimens, which then become the final product.

History

Human directed genetic manipulation began with the domestication of plants and animals through artificial selection in about 12,000 BC. Various techniques were developed to aid in breeding and selection. Hybridization was one way rapid changes in an organisms makeup could be introduced. Hybridization most likely first occurred when humans first grew similar, yet slightly different plants in close proximity. Some plants were able to be propagated by vegetative cloning. X-rays were first used to deliberately mutate plants in 1927. Between 1927 and 2017, more than 3,248 genetically mutated plant varieties had been produced using x-rays.

It wasn't until the mid 1800s that DNA and genes were discovered, which would form the basis of modern genetic manipulation. Genetic inheritance was first discovered by Gregor Mendel in 1865 following experiments crossing peas. In 1928 Frederick Griffith proved the existence of a "transforming principle" involved in inheritance, which was identified as DNA in 1944 by Oswald Avery, Colin MacLeod, and Maclyn McCarty. Frederick Sanger developed a method for sequencing DNA in 1977, greatly increasing the genetic information available to researchers.

As well as discovering how DNA works, tools had to be developed that allowed it to be manipulated. In 1970 Hamilton Smiths lab discovered restriction enzymes, enabling scientists to isolate genes from an organism's genome. DNA ligases, that join broken DNA together, had been discovered earlier in 1967 and by combining the two enzymes it was possible to "cut and paste" DNA sequences to create recombinant DNA. Plasmids, discovered in 1952, became important tools for transferring information between cells and replicating DNA sequences. Polymerase chain reaction (PCR), developed by Kary Mullis in 1983, allowed small sections of DNA to be amplified and aided identification and isolation of genetic material.

As well as manipulating the DNA, techniques had to be developed for its insertion (known as transformation) into an organism's genome. Griffiths experiment had already shown that some bacteria had the ability to naturally uptake and express foreign DNA. Artificial competence was induced in *Escherichia coli* in 1970 by treating them with calcium chloride solution ($CaCl_2$). Transformation using electroporation was developed in the late 1980s, increasing the efficiency and bacterial range. In 1907 a bacterium that caused plant tumors, *Agrobacterium tumefaciens*, had been discovered and in the early 1970s it was found that the bacteria inserted its DNA into plants using a Ti plasmid. By removing the genes in the plasmid that caused the tumor and adding in novel genes researchers were able to infect plants with *A. tumefaciens* and let the bacteria insert their chosen DNA into the genomes of the plants.

An important part of genetic engineering is to identify useful genes to transform into the genetically modified organism. The bacteria *Bacillus thuringiensis* was first discovered in 1901 as the causative agent in the death of silkworms. Due to these insecticidal properties the bacteria was used as an biological insecticide, commercially developed in 1938. The cry proteins were discovered to provide the insecticidal activity in 1956 and by the 1980s scientists had successfully cloned the gene coding for this protein and expressed it in plants. The gene that provides resistance to the glyphosate herbicide was found, after seven years searching, in the outflow pipe of a Monsanto roundup manufacturing facility. In animals the majority of genes used are growth hormone genes.

Libraries

Target genes can be cloned from a DNA segment after the creation of a DNA library. The libraries generally cover the organism's genome multiple times and its size will depend on how large the genome is.

Techniques

Gene Isolation

The DNA is first digested with a random digestion method, commonly by cutting the DNA with restriction enzymes (enzymes that cut DNA). A partial restriction digest cuts only some of the restriction sites, resulting in overlapping DNA fragment lengths. The DNA fragments are put into individual plasmid vectors and grown inside bacteria. Once in the bacteria the plasmid is copied as the bacteria divides. To determine if a useful gene is present on a particular fragment the bacterial library is screened for the desired phenotype. If the phenotype is detected then it is possible that the bacteria contains the target gene. If the gene does not have a detectable phenotype or a DNA library does not contain the correct gene, other methods can be used to isolate it. If the position of the gene can be determined using molecular markers then chromosome walking is one way to isolate the correct DNA fragment. If the gene expresses close homology to a known gene in another species, then it could be isolated by searching for genes in the library that closely match the known gene.

If the DNA sequence of the gene and the organism is known, restriction enzymes can cut the DNA on either side of the gene and gel electrophoresis can sort the fragments according to length. The DNA band at the correct size should contain the gene, where it can be excised from the gel. Polymerase chain reaction (PCR) can be used to amplify the gene, which can then be isolated through gel electrophoresis. It is also possible to synthesize the gene.

The gene to be inserted into the genetically modified organism must be combined with other genetic elements in order for it to work properly. The gene can also be modified at this stage for better expression or effectiveness. As well as the gene to be inserted most constructs contain a promoter and terminator region as well as a selectable marker gene. The promoter region initiates transcription of the gene and can be used to control the location and level of gene expression, while the terminator region ends transcription. The selectable marker, which in most cases confers antibiotic resistance to the organism it is expressed in, is needed to determine which cells are transformed with the new gene. The constructs are made using recombinant DNA techniques, such as restriction digests, ligations and molecular cloning.

Gene Targeting

Gene targeting uses homologous recombination to target desired changes to a specific endogenous gene. This tends to occur at a relatively low frequency in plants and animals and generally requires the use of selectable markers. The success of gene targeting can be enhanced with the use of engineered nucleases such as zinc finger nucleases, engineered homing endonucleases, transcription activator-like effector nuclease. or CRISPR. Engineered nucleases can also introduce mutations at endogenous genes that generate a gene knockout.

Transformation

About 1% of bacteria are naturally able to take up foreign DNA, but this ability can be induced in other bacteria. Stressing the bacteria with a heat shock or an electric shock can make the cell membrane permeable to DNA that may then incorporate into the genome or exist as extrachromosomal DNA. DNA is generally inserted into animal cells using microinjection, where it can be injected through the cells nuclear envelope directly into the nucleus or through the use of viral vectors. In plants the DNA is generally inserted using *Agrobacterium*-mediated recombination or biolistics.

In *Agrobacterium*-mediated recombination, the plasmid construct must also contain T-DNA. *Agrobacterium* naturally inserts DNA from a tumor-inducing plasmid into any susceptible plant that it infects, causing crown gall disease. The T-DNA region of this plasmid is responsible for insertion of the DNA. The DNA to be inserted is cloned into a binary vector that contains T-DNA and can be grown in both *E. coli* and *Agrobacterium*. Once the binary vector is constructed the plasmid is transformed into *Agrobacterium* containing no plasmids and plant cells are infected. The *Agrobacterium* naturally inserts the genetic material into the plant cells.

In biolistic transformation, particles of gold or tungsten are coated with DNA and then shot into young plant cells or plant embryos. Some genetic material enters the cells and transforms them. This method can be used on plants that are not susceptible to *Agrobacterium* infection and also allows transformation of plant plastids.

Another transformation method for plant and animal cells is electroporation, which involves subjecting cells to an electric shock, which can make the cell membrane permeable to plasmid DNA. In some cases the electroporated cells will incorporate the DNA. Due to the associated cell and DNA damage, the transformation efficiency of biolistics and electroporation is lower than with agrobacteria and microinjection.

Selection

Not all the organism's cells will be transformed with the new genetic material; typically a selectable marker is used to differentiate transformed from untransformed cells. Cells that have been successfully transformed with the DNA it will also contain the marker gene. By growing the cells in the presence of an antibiotic or chemical that selects or marks the cells expressing that gene, it is possible to separate modified from unmodified cells. Another screening method involves a DNA probe that sticks only to the inserted gene. Multiple strategies can remove the marker from the mature plant.

Regeneration

As often only a single cell is transformed with genetic material the modified organism must be grown from that single cell. Bacteria consists of a single cell and reproduce clonally, so regeneration is not necessary for them. In plants this is accomplished through the use of tissue culture. Each plant species has different requirements for successful regeneration. If successful, the technique produces an adult plant that contains the transgene in every cell.

In animals it is necessary to ensure that the inserted DNA is present in embryonic stem cells. Offspring can be screened for the gene. All offspring from the first generation will be heterozygous for the inserted gene and must be inbred to produce a homozygous specimen. Jurassic Park is a result of this.

Confirmation

The finding that a recombinant organism contains the inserted genes is not usually sufficient to ensure that they will be appropriately expressed in the intended tissues. To confirm the presence of the gene, PCR, Southern hybridization and DNA sequencing are employed to determine the chromosomal location and number of gene copies.

To assess gene expression, transcription, RNA processing patterns and expression and localization of protein product(s) must usually be assessed, using methods including northern hybridization, quantitative RT-PCR, Western blot, immunofluorescence and phenotypic analysis. When appropriate, the organism's offspring are studied to confirm that the trans-gene and associated phenotype are stably inherited.

In some cases further generations must be produced and confirmed, to ensure the absence of undesirable traits in the modified organism. For hybrid products such as maize, the modified organism is crossbred with other cultivars that possess required traits.

Gene Targeting

A chimeric mouse gene targeted for the agouti coat color gene, with its offspring

Gene targeting (also, replacement strategy based on homologous recombination) is a genetic technique that uses homologous recombination to change an endogenous gene. The method can be used to delete a gene, remove exons, add a gene, and introduce point mutations. Gene targeting can be permanent or conditional. Conditions can be a specific time during development / life of the organism or limitation to a specific tissue, for example. Gene targeting requires the creation of a specific vector for each gene of interest. However, it can be used for any gene, regardless of transcriptional activity or gene size.

Methods

Gene targeting methods are established for several model organisms and may vary depending on the species used. In general, a targeting construct made out of DNA is generated in bacteria. It typically contains part of the gene to be targeted, a reporter gene, and a (dominant) selectable marker.

To target genes in mice, this construct is then inserted into mouse embryonic stem cells in culture. After cells with the correct insertion have been selected, they can be used to contribute to a mouse's tissue via embryo injection. Finally, chimeric mice where the modified cells made up the reproductive organs are selected for via breeding. After this step the entire body of the mouse is based on the previously selected embryonic stem cell.

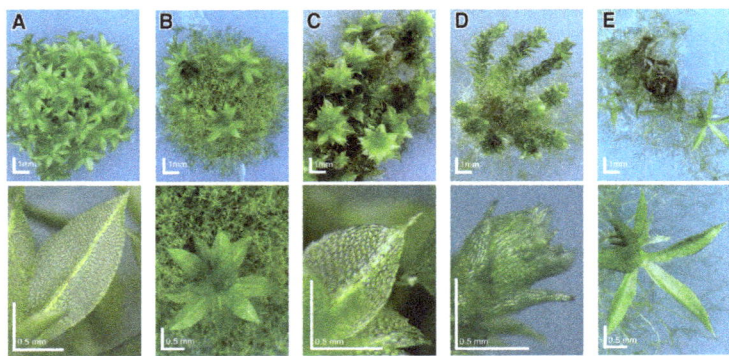

Wild-type Physcomitrella and knockout-mosses: Deviating phenotypes induced in gene-disruption library transformants. Physcomitrella wild-type and transformed plants were grown on minimal Knop medium to induce differentiation and development of gametophores. For each plant, an overview (upper row, scale bar corresponds to 1 mm) and a close-up (bottom row, scale bar equals 0.5 mm) is shown. A, Haploid wild-type moss plant completely covered with leafy gametophores and close-up of wild-type leaf. B-D, Different Mutants.

To target genes in moss, this construct is incubated together with freshly isolated protoplasts and with Polyethylene glycol. As mosses are haploid organisms, regenerating moss filaments (protonema) can directly be screened for gene targeting, either by treatment with antibiotics or with PCR. Unique among plants, this procedure for reverse genetics is as efficient as in yeast. Using modified procedures, gene targeting has also been successfully applied to cattle, sheep, swine, and many fungi.

The frequency of gene targeting can be significantly enhanced through the use of engineered endonucleases such as zinc finger nucleases, engineered homing endonucleases, and nucleases based on engineered TAL effectors. To date, this method has been applied to a number of species including Drosophila melanogaster, tobacco, corn, human cells, mice, and rats.

Comparison with Gene Trapping

Gene trapping is based on random insertion of a cassette while gene targeting targets a specific gene. Cassettes can be used for many different things while the flanking homology regions of gene targeting cassettes need to be adapted for each gene. This makes gene trapping more easily amenable for large scale projects than targeting. On the other hand, gene targeting can be used for genes with low transcriptions that would go undetected in a trap screen. Also, the probability of trapping increases with intron size. For gene targeting these compact genes are just as easily altered.

Applications

Gene targeting has been widely used to study human genetic diseases by removing ("knocking out"), or adding ("knocking in"), specific mutations of interest to a variety of models. Previously used to engineer rat cell models, advances in gene targeting technologies are enabling the creation of a new wave of isogenic human disease models. These models are the most accurate in-vitro models available to researchers to date, and are facilitating the development of new personalized drugs and diagnostics, particularly in the field of cancer.

2007 Nobel Prize

Mario R. Capecchi, Martin J. Evans and Oliver Smithies were declared laureates of the 2007 Nobel Prize in Physiology or Medicine for their work on "principles for introducing specific gene modifications in mice by the use of embryonic stem cells", or gene targeting.

Gene Knockin

In molecular cloning and biology, a knock-in (or gene knock-in) refers to a genetic engineering method that involves the one-for-one substitution of DNA sequence information with a wild-type copy in a genetic locus or the insertion of sequence information not found within the locus. Typically, this is done in mice since the technology for this process is more refined and there is a high degree of shared sequence complexity between mice and humans. The difference between knock-in technology and traditional transgenic techniques is that a knock-in involves a gene inserted into a specific locus, and is thus a "targeted" insertion.

A common use of knock-in technology is for the creation of disease models. It is a technique by which scientific investigators may study the function of the regulatory machinery (e.g. promoters) that governs the expression of the natural gene being replaced. This is accomplished by observing the new phenotype of the organism in question. The BACs and YACs are used in this case so that large fragments can be transferred.

Technique

Gene knockin originated as a slight modification of the original knockout technique developed by Martin Evans, Oliver Smithies, and Mario Capecchi. Traditionally, knockin techniques have relied on homologous recombination to drive targeted gene replacement, although other methods using a transposon-mediated system to insert the target gene have been developed. The use of *loxP* flanking sites that become excised upon expression of Cre recombinase with gene vectors is an example of this. Embryonic stem cells with the modification of interest are then implanted into a viable blastocyst, which will grow into a mature chimeric mouse with some cells having the original blastocyst cell genetic information and other cells having the modifications introduced to the embryonic stem cells. Subsequent offspring of the chimeric mouse will then have the gene knockin.

Gene knockin has allowed, for the first time, hypothesis-driven studies on gene modifications and resultant phenotypes. Mutations in the human p53 gene, for example, can be induced by exposure to benzo(a)pyrene and the mutated copy of the p53 gene can be inserted into mouse genomes. Lung tumors observed in the knockin mice offer support for the hypothesis of BaP's carcinogenicity. More recent developments in knockin technique have allowed for pigs to have a gene for green fluorescent protein inserted with a CRISPR/Cas9 system, which allows for much more accurate and successful gene insertions. The speed of CRISPR/Cas9-mediated gene knockin also allows for biallelic modifications to some genes to be generated and the phenotype in mice observed in a single generation, an unprecedented timeframe.

Versus Gene Knockout

Knockin technology is different from knockout technology in that knockout technology aims to either delete part of the DNA sequence or insert irrelevant DNA sequence information to disrupt the expression of a specific genetic locus. Gene knockin technology, on the other hand, alters the genetic locus of interest via a one-for-one substitution of DNA sequence information or by the addition of sequence information that is not found on said genetic locus. A gene knockin therefore can be seen as a gain of function mutation and a gene knockout a loss of function mutation, but a gene knockin may also involve the substitution of a functional gene locus for a mutant phenotype that results in some loss of function.

Potential Applications

Because of the success of gene knockin methods thus far, many clinical applications can be envisioned. Knockin of sections of the human immunoglobulin gene into mice has already been shown to allow them to produce humanized antibodies that are therapeutically useful. It should be possible to modify stem cells in humans to restore targeted gene function in certain tissues, for example possibly correcting the mutant gamma-chain gene of the IL-2 receptor in hematopoietic stem cells to restore lymphocyte development in people with X-linked severe combined immunodeficiency.

Limitations

While gene knockin technology has proven to be a powerful technique for the generation of models of human disease and insight into proteins *in vivo*, numerous limitations still exist. Many of these are shared with the limitations of knockout technology. First, combinations of knockin genes lead to growing complexity in the interactions that inserted genes and their products have with other sections of the genome and can therefore lead to more side effects and difficult-to-explain phenotypes. Also, only a few loci, such as the ROSA26 locus have been characterized well enough where they can be used for conditional gene knockins, making combinations of reporter and transgenes in the same locus problematic. The biggest disadvantage of using gene knockin for human disease model generation is that mouse physiology is not identical to that of humans and human orthologs of proteins expressed in mice will often not wholly reflect the role of a gene in human pathology. This can be seen in mice produced with the ΔF508 fibrosis mutation in the CFTR gene, which accounts for more than 70% of the mutations in this gene for the human population and leads to cystic fibrosis. While ΔF508 CF mice do exhibit the processing defects characteristic of the human mutation, they do not display the pulmonary pathophysiological changes seen in humans and carry virtually no lung phenotype. Such problems could be ameliorated by the use of a variety of animal models, and pig models (pig lungs share many biochemical and physiological similarities with human lungs) have been generated in an attempt to better explain the activity of the ΔF508 mutation.

Gene Drive

In genetics, gene drive is a technique that promotes the inheritance of a particular gene to increase its prevalence in a population. The term is also used for specific genetic elements (i.e. a piece of DNA) that can implement the technique.

Applications of gene drive include preventing the spread of insects that carry pathogens (in particular, mosquitoes that transmit malaria, dengue, and zika pathogens), to control invasive species, or to eliminate herbicide or pesticide resistance. The technique can be used for adding, disrupting, or modifying genes, such as to cause a crash in the populations of a disease vector by reducing their reproductive capacity.

Several molecular mechanisms can mediate gene drive. Naturally occurring gene drive mechanisms arise when alleles evolve molecular mechanisms that give them a transmission chance greater than the normal 50%. Synthetic genetic modules with similar properties have been developed as a technique for genome editing of laboratory populations. This entry focuses on endonuclease-based gene drive, the most versatile and actively developing molecular backend for synthetic gene drives. Since gene drives function only in sexually reproducing species, they cannot be used to engineer populations of viruses or bacteria.

Because it is a way to artificially bias inheritance of desired genes, gene drive constitutes a major change in biotechnology. The potential impact of releasing gene drives in the wild raises major bioethical concerns regarding their development and management.

Mechanism

In sexually-reproducing species, most genes are present in two copies (which can be different alleles or not), each of which has 50% chance of being inherited. For a particular allele to spread through a large population, it must increase the fitness of each individual. However, some alleles have evolved molecular mechanisms that confer them greater transmission chance than the normal 50%. This allows them to spread through a population even if they reduce the fitness of each individual organism. By similarly biasing the inheritance of particular altered genes, synthetic gene drives might be used to spread alterations through wild populations.

Molecular Mechanisms

At the molecular level, endonuclease gene drives work by cutting chromosomes that do not encode the drive at a specific site, inducing the cell to repair the damage by copying the drive sequence onto the damaged chromosome. This is derived from genome editing techniques and similarly relies on the fact that double strand breaks are most frequently repaired by homologous recombination if a template is present, and less often by non-homologous end joining. The cell then has two copies of the drive sequence. To achieve this behavior, endonuclease gene drives consist of two nested elements:

- either a homing endonuclease or a RNA-guided endonuclease (e.g. Cas9 or Cpf1) and its guide RNA, that cuts the target sequence in recipient cells

- a template sequence used by the DNA repair machinery after the target sequence is cut. To achieve the self propagating nature of gene drives, this repair template contains at least the endonuclease sequence. Because the template must be used to repair a double-strand break at the cutting site, its sides are homologous to the sequences that are adjacent to the cutting site in the host genome. By targeting the gene drive to a gene coding sequence, this gene will be inactivated; additional sequences can be introduced in the gene drive to encode new functions.

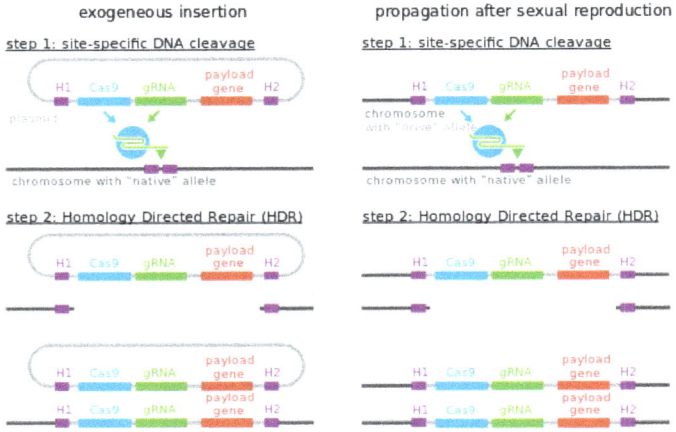

Molecular mechanism of gene drive.

As a result, the gene drive insertion in the genome will re-occur in each organism that inherits one copy of the modification and one copy of the wild-type gene. If the gene drive is already present in the egg cell (e.g. when received from one parent), all the gametes of the individual will carry the gene drive (instead of 50% in the case of a normal gene).

Spreading in the Population

Since it can never more than double in frequency with each generation, a gene drive introduced in a single individual typically requires dozens of generations to affect a substantial fraction of a population. Alternatively, releasing drive-containing organisms in sufficient numbers can affect the rest within a few generations; for instance, by introducing it in every thousandth individual, it takes only 12–15 generations to be present in all individuals. Whether a gene drive will ultimately become fixed in a population and at which speed depends on its effect on individuals fitness, on the rate of allele conversion, and on the population structure. In a well mixed population and with realistic allele conversion frequencies (≈90%), population genetics predicts that gene drives get fixed for selection coefficient smaller than 0.3; in other words, gene drives can be used not only to spread beneficial genetic modifications, but also detrimental ones as long the reproductive success is not reduced by more than 30%. This is a great contrast with normal genes, which can only spread in large populations if they are beneficial.

Applications and Technical Limitations

Applications

Gene drives have two main classes of applications, which, although based on the same technology, have implications of different significance:

- introduce a genetic modification in laboratory populations; once a strain or a line carrying the gene drive has been produced, the drive can be passed on to any other line simply by mating. Here the gene drive is used to achieve much more easily a task that could be accomplished with other techniques. It requires reinforced confinement of the lab populations to prevent an accidental release of the gene drive into the wild.

- introduce a genetic modification in wild populations. In contrast with the former, gene drives here constitute a major development and open the door to previously unattainable changes. This raises major ethical issues.

Because of the unprecedented potential of gene drives, safeguard mechanisms have been proposed and tested.

Technical Limitations

Because gene drives propagate by replacing other alleles that contain a cutting site and the corresponding homologies, their application is limited to sexually reproducing species (because they are diploid and alleles are mixed at each generation). As a side effect, inbreeding could in principle be selected as an escape mechanism, but the extent to which this can happen in practice is difficult to evaluate.

Due to the number of generations required for a gene drive to spread in an entire population, it may require under a year for some invertebrates, but centuries for organisms with years-long intervals between birth and sexual maturity, such as humans. Hence this technology is of most use in fast-reproducing species.

Issues

Issues that researchers have highlighted include:

- Mutations—It is possible that a mutation could happen mid-drive, which has the potential to allow unwanted traits to "ride along" on the spreading drive.

- Escape—Cross-breeding or gene flow potentially allow a drive to move beyond its target population.

- Ecological impacts—Even when new traits' direct impact on a target is understood, the drive may have side effects on the surroundings.

There are bioethics concerns as well, as the gene drive is a very powerful tool.

In December 2015, scientists of major world academies called for a moratorium on inheritable human genome edits that would be passed on in pregnancies, including those related to CRIS-

PR-Cas9 technologies, but supported continued basic research and gene editing that would not affect future generations. In February 2016, British scientists were given permission by regulators to genetically modify human embryos by using CRISPR-Cas9 and related techniques on condition that the embryos were destroyed in seven days. In June 2016, the US National Academies of Sciences, Engineering, and Medicine released a report on their "Recommendations for Responsible Conduct" of gene drives.

History

Austin Burt, an evolutionary geneticist at Imperial College London, first outlined the possibility of building gene drives based on natural "selfish" homing endonuclease genes in 2003. Researchers had already shown that these "selfish" genes could spread rapidly through successive generations. Burt suggested that gene drives might be used to prevent a mosquito population from transmitting the malaria parasite or crash a mosquito population. Gene drives based on homing endonucleases have been demonstrated in the laboratory in transgenic populations of mosquitoes and fruit flies. However, homing endonucleases are sequence-specific. Since altering their specificity to target other sequences of interest remains a major challenge, the possible applications of gene drive remained limited until the discovery of CRISPR and the associated RNA-guided endonucleases such as Cas9 and Cpf1.

In August 2016 the U.S. Food And Drug Administration (FDA) issued a "Finding of No Significant Impact" (FONSI) to biotech company Oxitec's plan to release genetically modified male *Aedes aegypti* (mosquitoes) into the Florida Keys. The intent was to stop the spread of mosquito-borne diseases, including Zika. The modification adds a gene that kills their offspring before they reach reproductive age. Oxitec still needs approval from the Florida Keys Mosquito Control District before releasing any insects.

The Bill and Melinda Gates Foundation has invested $75 million in gene drive technology. The foundation originally estimated the technology to be ready for field use by 2029 somewhere in Africa. However, Gates in 2016 changed this estimate to instead happen some time within the next two years.

CRISPR/Cas9

CRISPR/Cas9 is a DNA cutting method that has made genetic engineering faster, easier, and more efficient since 2013. The approach involves expressing the RNA-guided Cas9 endonuclease along with guide RNAs directing it to a particular sequence to be edited. When Cas9 cuts the target sequence, the cell often repairs the damage by replacing the original sequence with homologous DNA. By introducing an additional template with appropriate homologies, Cas9 can be used to delete, add, or modify genes in an unprecedentedly simple manner. As of 2014, it had successfully been tested in cells of 20 species, including humans. In many of these species, the edits modified their germline, allowing them to be inherited.

In 2014 Kevin Esvelt and coworkers first suggested that CRISPR/Cas9 might be used to build endonuclease gene drives. In 2015 researchers published successful engineering of CRISPR-based gene drives in *Saccharomyces, Drosophila* and mosquitoes. All four studies demonstrated extremely efficient inheritance distortion over successive generations, with one study demonstrating the spread of a gene drive into naïve laboratory populations. Drive-resistant alleles are expected to arise for each of the described gene drives, however this can be delayed or prevented by targeting highly conserved sites at which resistance is expected to have a severe fitness cost.

Because of CRISPR/Cas9's targeting flexibility, the derived gene drives could theoretically be used to engineer almost any trait. Unlike previous designs, they could be tailored to block the evolution of drive resistance in the target population by targeting multiple sequences within appropriate genes. CRISPR/Cas9 could also permit a variety of gene drive architectures intended to control rather than crash populations. Noticeably, RNA-guided gene drives could be designed with other RNA-guided endonuclease such as CRISPR/Cpf1.

Applications to Wild Populations

Disease Vector Species

One possible application is to genetically modify mosquitoes and other disease vectors so they cannot transmit diseases such as malaria and dengue fever. In June 2014, the World Health Organization (WHO) Special Programme for Research and Training in Tropical Diseases issued guidelines for evaluating genetically modified mosquitoes. In 2013 the European Food Safety Authority issued a protocol for environmental assessments of all genetically modified organisms. Researchers believe that by applying the new technique to just 1% of the wild population of mosquitoes, they can eradicate malaria within a year.

Invasive Species

Gene drive could be used to eliminate invasive species and has, for example, been proposed as a way to eliminate invasive species in New Zealand; in response others have raised the standard objections to the use of gene drive.

Predator Free 2050

In July 2016, New Zealand's prime announced the Predator Free 2050 project, a government program to completely eliminate eight invasive mammalian predator species (various rats, short-tailed weasels, and possums) from the New Zealand mainland by 2050. In January 2017 it was announced that gene drive technology would be used in the effort.

In 2017, two groups, one in Australia and another in Texas, released preliminary research into creating 'daughterless mice', using gene drives in mammals for the first time; these 'daughterless mice' are considered a breakthrough and particularly useful for New Zealand and other islands overrun with invasive mammals.

Gene Knockout

A gene knockout (abbreviation: KO) is a genetic technique in which one of an organism's genes is made inoperative ("knocked out" of the organism). Also known as knockout organisms or simply knockouts, they are used in learning about a gene that has been sequenced, but which has an unknown or incompletely known function. Researchers draw inferences from the difference between the knockout organism and normal individuals.

The term also refers to the process of creating such an organism, as in "knocking out" a gene. The technique is essentially the opposite of a gene knockin. Knocking out two genes simultaneously in an organism is known as a double knockout (DKO). Similarly the terms triple knockout (TKO) and quadruple knockouts (QKO) are used to describe three or four knocked out genes, respectively.

Knockout is accomplished through a combination of techniques, beginning in the test tube with a plasmid, a bacterial artificial chromosome or other DNA construct, and proceeding to cell culture. Individual cells are genetically transfected with the DNA construct. Often the goal is to create a transgenic animal that has the altered gene. If so, embryonic stem cells are genetically transformed and inserted into early embryos. Resulting animals with the genetic change in their germline cells can then often pass the gene knockout to future generations.

To create knockout moss, transfection of protoplasts is the preferred method. Such transformed *Physcomitrella*-protoplasts directly regenerate into fertile moss plants. Eight weeks after transfection, the plants can be screened for gene targeting via PCR.

The construct is engineered to recombine with the target gene, which is accomplished by incorporating sequences from the gene itself into the construct. Recombination then occurs in the region of that sequence within the gene, resulting in the insertion of a foreign sequence to disrupt the gene. With its sequence interrupted, the altered gene in most cases will be translated into a nonfunctional protein, if it is translated at all.

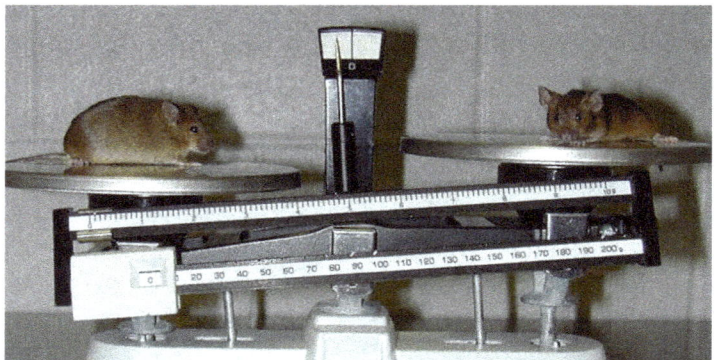

A knockout mouse (left) that is a model of obesity, compared with a normal mouse.

A conditional knockout allows gene deletion in a tissue or time specific manner. This is done by introducing short sequences called loxP sites around the gene. These sequences will be introduced into the germ-line via the same mechanism as a knock-out. This germ-line can then be crossed to another germline containing Cre-recombinase which is a viral enzyme that can recognize these sequences, recombines them and deletes the gene flanked by these sites.

Because the desired type of DNA recombination is a rare event in the case of most cells and most constructs, the foreign sequence chosen for insertion usually includes a reporter. This enables easy selection of cells or individuals in which knockout was successful. Sometimes the DNA construct inserts into a chromosome without the desired homologous recombination with the target gene.

To eliminate such cells, the DNA construct often contains a second region of DNA that allows such cells to be identified and discarded.

In diploid organisms, which contain two alleles for most genes, and may as well contain several related genes that collaborate in the same role, additional rounds of transformation and selection are performed until every targeted gene is knocked out. Selective breeding may be required to produce homozygous knockout animals.

Gene knockin is similar to gene knockout, but it replaces a gene with another instead of deleting it.

Use

Knockouts are primarily used to understand the role of a specific gene or DNA region by comparing the knockout organism to a wildtype with a similar genetic background.

Knockouts organisms are also used as screening tools in the development of drugs, to target specific biological processes or deficiencies by using a specific knockout, or to understand the mechanism of action of a drug by using a library of knockout organisms spanning the entire genome, such as in *Saccharomyces cerevisiae*.

Gene Knockdown

Gene knockdown is an experimental technique by which the expression of one or more of an organism's genes are reduced. The reduction can occur either through genetic modification or by treatment with a reagent such as a short DNA or RNA oligonucleotide that has a sequence complementary to either gene or an mRNA transcript.

Versus Transient Knockdown

If genetic modification of DNA is done, the result is called "knockdown organism." If the change in gene expression is caused by an oligonucleotide binding to an mRNA or temporarily binding to a gene, this leads to a temporary change in gene expression that does not modify the chromosomal DNA, and the result is referred to as a "transient knockdown".

In a transient knockdown, the binding of this oligonucleotide to the active gene or its transcripts causes decreased expression through a variety of processes. Binding can occur either through the blocking of transcription (in the case of gene-binding), the degradation of the mRNA transcript (e.g. by small interfering RNA (siRNA)) or RNase-H dependent antisense), or through the blocking of either mRNA translation, pre-mRNA splicing sites, or nuclease cleavage sites used for maturation of other functional RNAs, including miRNA (e.g. by morpholino oligos or other RNase-H independent antisense).

The most direct use of transient knockdowns is for learning about a gene that has been sequenced, but has an unknown or incompletely known function. This experimental approach is known as reverse genetics. Researchers draw inferences from how the knockdown differs from individuals in which the gene of interest is operational. Transient knockdowns are often used in developmental biology because oligos can be injected into single-celled zygotes and will be present in the daughter cells of the injected cell through embryonic development.

RNA Interference

RNA interference (RNAi) is a means of silencing genes by way of mRNA degradation. Gene knockdown by this method is achieved by introducing small double-stranded interfering RNAs (siRNA) into the cytoplasm. Small interfering RNAs can originate from inside the cell or can be exogenously introduced into the cell. Once introduced into the cell, exogenous siRNAs are processed by the RNA-induced silencing complex (RISC). The siRNA is complementary to the target mRNA to be silenced, and the RISC uses the siRNA as a template for locating the target mRNA. After the RISC localizes to the target mRNA, the RNA is cleaved by a ribonuclease.

RNAi is widely used as a laboratory technique for genetic functional analysis. RNAi in organisms such as *C. elegans* and *Drosophila melanogaster* provides a quick and inexpensive means of investigating gene function. In *C. elegans* research, the availability of tools such as the Ahringer RNAi Library give laboratories a way of testing many genes in a variety of experimental backgrounds. Insights gained from experimental RNAi use may be useful in identifying potential therapeutic targets, drug development, or other applications. RNA interference is a very useful research tool, allowing investigators to carry out large genetic screens in an effort to identify targets for further research related to a particular pathway, drug, or phenotype.

CRISPRs

A different means of silencing exogenous DNA that has been discovered in prokaryotes is a mechanism involving loci called 'Clustered Regularly Interspaced Short Palindromic Repeats', or CRISPRs. Proteins called 'CRISPR-associated genes' (cas genes) encode cellular machinery that cuts exogenous DNA into small fragments and inserts them into a CRISPR repeat locus. When this CRISPR region of DNA is expressed by the cell, the small RNAs produced from the exogenous DNA inserts serve as a template sequence that other Cas proteins use to silence this same exogenous sequence. The transcripts of the short exogenous sequences are used as a guide to silence these foreign DNA when they are present in the cell. This serves as a kind of acquired immunity, and this process is like a prokaryotic RNA interference mechanism. The CRISPR repeats are conserved amongst many species and have been demonstrated to be usable in human cells, bacteria, *C. elegans*, zebrafish, and other organisms for effective genome manipulation. The use of CRISPRs as a versatile research tool can be illustrated by many studies making use of it to generate organisms with genome alterations.

TALENs

Another technology made possible by prokaryotic genome manipulation is the use of transcription activator-like effector nucleases (TALENs) to target specific genes. TALENs are nucleases that have two important functional components: a DNA binding domain and a DNA cleaving domain. The DNA binding domain is a sequence-specific transcription activator-like effector sequence while the DNA cleaving domain originates from a bacterial endonuclease and is non-specific. TALENs can be designed to cleave a sequence specified by the sequence of the transcription activator-like effector portion of the construct. Once designed, a TALEN is introduced into a cell as a plasmid or mRNA. The TALEN is expressed, localizes to its target sequence, and cleaves a specific site. After cleavage of the target DNA sequence by the TALEN, the cell uses non-homologous end joining as a DNA repair mechanism to correct the cleavage. The cell's attempt at repairing the cleaved sequence can render the encoded protein non-functional, as this repair mechanism introduces insertion or deletion errors at the repaired site.

Commercialization

So far, knockdown organisms with permanent alterations in their DNA have been engineered chiefly for research purposes. Also known simply as knockdowns, these organisms are most commonly used for reverse genetics, especially in species such as mice or rats for which transient knockdown technologies cannot easily be applied.

There are several companies that offer commercial services related to gene knockdown treatments.

Electroporation

Electroporation, or electropermeabilization, is a microbiology technique in which an electrical field is applied to cells in order to increase the permeability of the cell membrane, allowing chemicals, drugs, or DNA to be introduced into the cell. In microbiology, the process of electroporation is often used to transform bacteria, yeast, or plant protoplasts by introducing new coding DNA. If bacteria and plasmids are mixed together, the plasmids can be transferred into the bacteria after electroporation, though depending on what is being transferred cell-penetrating peptides or CellSqueeze could also be used. Electroporation works by passing thousands of volts across a distance of one to two millimeters of suspended cells in an electroporation cuvette (1.0 − 1.5 kV, 250 − 750V/cm). Afterwards, the cells have to be handled carefully until they have had a chance to divide, producing new cells that contain reproduced plasmids. This process is approximately ten times more effective than chemical transformation.

Electroporation is also highly efficient for the introduction of foreign genes into tissue culture cells, especially mammalian cells. For example, it is used in the process of producing knockout mice, as well as in tumor treatment, gene therapy, and cell-based therapy. The process of introducing foreign DNA into eukaryotic cells is known as transfection. Electroporation is highly effective for transfecting cells in suspension using electroporation cuvettes. Electroporation has proven efficient for use on tissues in vivo, for in utero applications as well as in ovo transfection. Adherent cells can also be transfected using electroporation, providing researchers with an alternative to trypsinizing their cells prior to transfection. One downside to electroporation, however, is that after the process the gene expression of over 7,000 genes can be affected. This can cause problems in studies where gene expression has to be controlled to ensure accurate and precise results.

Laboratory Practice

Electroporation is performed with electroporators, purpose-built appliances which create an electrostatic field in a cell solution. The cell suspension is pipetted into a glass or plastic cuvette which has two aluminum electrodes on its sides. For bacterial electroporation, typically a suspension of around 50 microliters is used. Prior to electroporation, this suspension of bacteria is mixed with the plasmid to be transformed. The mixture is pipetted into the cuvette, the voltage and capacitance are set, and the cuvette is inserted into the electroporator. The process requires direct contact between the electrodes and the suspension. Immediately after electroporation, one milliliter of liquid medium is added to the bacteria (in the cuvette or in an Eppendorf tube), and the tube is incubated at the bacteria's optimal temperature for an hour or more to allow recovery of the cells and expression of the plasmid, followed by bacterial culture on agar plates.

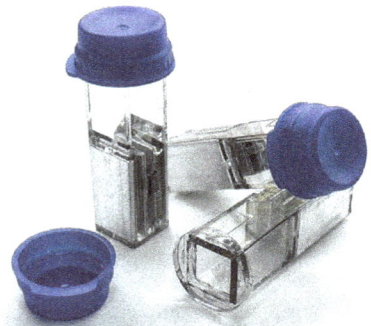

Cuvettes for electroporation. These are plastic with aluminium electrodes and a blue lid.
They hold a maximum of 400 μl.

The success of the electroporation depends greatly on the purity of the plasmid solution, especially on its salt content. Solutions with high salt concentrations might cause an electrical discharge (known as arcing), which often reduces the viability of the bacteria. For a further detailed investigation of the process, more attention should be paid to the output impedance of the porator device and the input impedance of the cells suspension (e.g. salt content).

Since the cell membrane is not able to pass current (except in ion channels), it acts as an electrical capacitor. Subjecting membranes to a high-voltage electric field results in their temporary breakdown, resulting in pores that are large enough to allow macromolecules (such as DNA) to enter or leave the cell.

Medical Applications

The first group to look at electroporation for medical applications was led by Lluis M Mir at the Institute Gustave Roussy. In this case, they looked at the use of reversible electroporation in conjuction with impermeable macromolecules. The first research looking at how nanosecond pulses might be used on human cells was conducted by researchers at Eastern Virginia Medical School and Old Dominion University, and published in 2003.

With regards to irreversible electroporation, the first successful treatment of malignant cutaneous tumors implanted in mice was completed in 2007 by a group of scientists who achieved complete tumor ablation in 12 out of 13 mice. They accomplished this by sending 80 pulses of 100 microseconds at 0.3 Hz with an electrical field magnitude of 2500 V/cm to treat the cutaneous tumors.

A higher voltage of electroporation was found in pigs to irreversibly destroy target cells within a narrow range while leaving neighboring cells unaffected, and thus represents a promising new treatment for cancer, heart disease and other disease states that require removal of tissue. Irreversible electroporation (IRE) has since proven effective in treating human cancer, with surgeons at Johns Hopkins and other institutions now using the technology to treat pancreatic cancer previously thought to be unresectable.

A recent technique called non-thermal irreversible electroporation (N-TIRE) has proven successful in treating many different types of tumors and other unwanted tissue. This procedure is done using small electrodes (about 1mm in diameter), placed either inside or surrounding the target tissue to apply short, repetitive bursts of electricity at a predetermined voltage and frequency. These bursts of electricity increase the resting transmembrane potential (TMP), so that nanopores form in the plasma membrane. When the electricity applied to the tissue is above the electric field threshold of the target tissue, the cells become permanently permeable from the formation of nanopores. As a result, the cells are unable to repair the damage and die due to a loss of homeostasis. N-TIRE is unique to other tumor ablation techniques in that it does not create thermal damage to the tissue around it.

Contrastingly, reversible electroporation occurs when the electricity applied with the electrodes is below the electric field threshold of the target tissue. Because the electricity applied is below the cells' threshold, it allows the cells to repair their phospholipid bilayer and continue on with their normal cell functions. Reversible electroporation is typically done with treatments that involve getting a drug or gene (or other molecule that is not normally permeable to the cell membrane) into the cell. Not all tissue has the same electric field threshold; therefore careful calculations need to be made prior to a treatment to ensure safety and efficacy.

One major advantage of using N-TIRE is that, when done correctly according to careful calculations, it only affects the target tissue. Proteins, the extracellular matrix, and critical structures such as blood vessels and nerves are all unaffected and left healthy by this treatment. This allows for a quicker recovery, and facilitates a more rapid replacement of dead tumor cells with healthy cells.

Before doing the procedure, scientists must carefully calculate exactly what needs to be done, and treat each patient on an individual case-by-case basis. To do this, imaging technology such as CT scans and MRI's are commonly used to create a 3D image of the tumor. From this information, they can approximate the volume of the tumor and decide on the best course of action including the insertion site of electrodes, the angle they are inserted in, the voltage needed, and more, using software technology. Often, a CT machine will be used to help with the placement of electrodes during the procedure, particularly when the electrodes are being used to treat tumors in the brain.

The entire procedure is very quick, typically taking about five minutes. The success rate of these procedures is high and is very promising for future treatment in humans. One disadvantage to using N-TIRE is that the electricity delivered from the electrodes can stimulate muscle cells to contract, which could have lethal consequences depending on the situation. Therefore, a paralytic agent must be used when performing the procedure. The paralytic agents that have been used in such research are successful; however, there is always some risk, albeit slight, when using anesthetics.

A more recent technique has been developed called high-frequency irreversible electroporation (H-FIRE). This technique uses electrodes to apply bipolar bursts of electricity at a high frequency, as opposed to unipolar bursts of electricity at a low frequency. This type of procedure has the same tumor ablation success as N-TIRE. However, it has one distinct advantage, H-FIRE does not cause muscle contraction in the patient and therefore there is no need for a paralytic agent.

Drug and Gene Delivery

Electroporation can also be used to help deliver drugs or genes into the cell by applying short and intense electric pulses that transiently permeabilize cell membrane, thus allowing transport of molecules otherwise not transported through a cellular membrane. This procedure is referred to as electrochemotherapy when the molecules to be transported are chemotherapeutic agents or gene electrotransfer when the molecule to be transported is DNA. Scientists from Karolinska Institutet and the University of Oxford use electroporation of exosomes to deliver siRNAs, antisense oligonucleotides, chemotherapeutic agents and proteins specifically to neurons after inject them systemically (in blood). Because these exosomes are able to cross the blood brain barrier this protocol could solve the issue of poor delivery of medications to the central nervous system and cure Alzheimer's, Parkinson's Disease and brain cancer among other diseases.

Physical Mechanism

Electroporation allows cellular introduction of large highly charged molecules such as DNA which would never passively diffuse across the hydrophobic bilayer core. This phenomenon indicates that the mechanism is the creation of nm-scale water-filled holes in the membrane. Although electroporation and dielectric breakdown both result from application of an electric field, the mechanisms involved are fundamentally different. In dielectric breakdown the barrier material is ionized, creating a conductive pathway. The material alteration is thus chemical in nature. In contrast, during electroporation the lipid molecules are not chemically altered but simply shift position, opening up a pore which acts as the conductive pathway through the bilayer as it is filled with water.

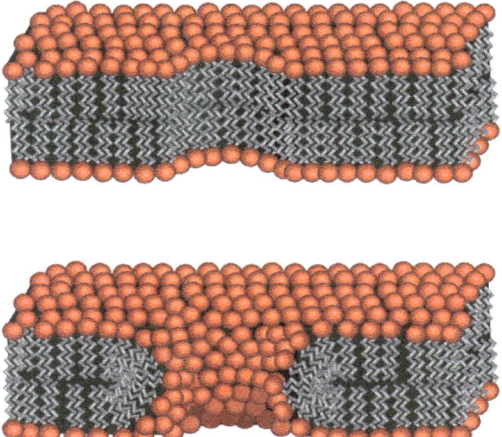

Schematic showing the theoretical arrangement of lipids in a hydrophobic pore (top) and a hydrophilic pore (bottom).

Electroporation is a dynamic phenomenon that depends on the local transmembrane voltage at each point on the cell membrane. It is generally accepted that for a given pulse duration and shape, a specific transmembrane voltage threshold exists for the manifestation of the electroporation phenomenon (from 0.5 V to 1 V). This leads to the definition of an electric field magnitude threshold for electroporation (E_{th}). That is, only the cells within areas where $E \geqq E_{th}$ are electroporated. If a second threshold (E_{ir}) is reached or surpassed, electroporation will compromise the viability of the cells, *i.e.*, irreversible electroporation (IRE).

Electroporation is a multi-step process with several distinct phases. First, a short electrical pulse must be applied. Typical parameters would be 300–400 mV for < 1 ms across the membrane (the voltages used in cell experiments are typically much larger because they are being applied across large distances to the bulk solution so the resulting field across the actual membrane is only a small fraction of the applied bias). Upon application of this potential the membrane charges like a capacitor through the migration of ions from the surrounding solution. Once the critical field is achieved there is a rapid localized rearrangement in lipid morphology. The resulting structure is believed to be a "pre-pore" since it is not electrically conductive but leads rapidly to the creation of a conductive pore. Evidence for the existence of such pre-pores comes mostly from the "flickering" of pores, which suggests a transition between conductive and insulating states. It has been suggested that these pre-pores are small (~3 Å) hydrophobic defects. If this theory is correct, then the transition to a conductive state could be explained by a rearrangement at the pore edge, in which the lipid heads fold over to create a hydrophilic interface. Finally, these conductive pores can either heal, resealing the bilayer or expand, eventually rupturing it. The resultant fate depends on whether the critical defect size was exceeded which in turn depends on the applied field, local mechanical stress and bilayer edge energy.

References

- Mandel, Morton; Higa, Akiko (1970). "Calcium-dependent bacteriophage DNA infection". Journal of Molecular Biology. 53 (1): 159–162. PMID 4922220. doi:10.1016/0022-2836(70)90051-3

- Corinne A. Michels (2002). "7". Genetic Techniques for Biological Research: A Case Study Approach. John Wiley & Sons. pp. 85–88. ISBN 0-471-89919-4

- S.C. Ekker (2008). "Zinc finger-based knockout punches for zebrafish genes". Zebrafish. 5 (2): 1121–3. PMC 2849655. PMID 18554175. doi:10.1089/zeb.2008.9988

- "EFSA - Guidance of the GMO Panel: Guidance Document on the ERA of GM animals". EFSA Journal. 11 (5): 3200. 2013. doi:10.2903/j.efsa.2013.3200. Retrieved 2014-07-18

- Graham Head; Hull, Roger H; Tzotzos, George T. (2009). Genetically Modified Plants: Assessing Safety and Managing Risk. London: Academic Pr. p. 244. ISBN 0-12-374106-8

- "U.S. researchers call for greater oversight of powerful genetic technology | Science/AAAS | News". News.sciencemag.org. Retrieved 2014-07-18

- Bibikova, M.; Beumer, K.; Trautman, J.; Carroll, D. (2003). "Enhancing Gene Targeting with Designed Zinc Finger Nucleases". Science. 300 (5620): 764. PMID 12730594. doi:10.1126/science.1079512

- Pratt, AJ (2009). "The RNA-induced silencing complex: a versatile gene-silencing machine". Journal of Biological Chemistry. 284: 17897–901. PMC 2709356. PMID 19342379. doi:10.1074/jbc.R900012200

- Gibson, Greg (2009). A Primer Of Genome Science 3rd ed. Sunderland, Massachusetts: Sinauer. pp. 301–302. ISBN 978-0-87893-236-8

- Huffaker, Sandy (9 December 2015). "Geneticists vote to allow gene editing of human embryos". New Scientist. Retrieved 18 March 2016

- Ghadakzadeh, S; Mekhail, M; Aoude, A; Hamdy, R; Tabrizian, M (2016). "Small Players Ruling the Hard Game: siRNA in Bone Regeneration". Journal of Bone and Mineral Research. 31 (3): 475–487. doi:10.1002/jbmr.2816

- Gantz, V. M.; Bier, E. (2015). "The mutagenic chain reaction: A method for converting heterozygous to homozygous mutations". Science. 348 (6233): 442–444. PMC 4687737. PMID 25908821. doi:10.1126/science.aaa5945

- Garcia, P A; Neal, Robert E; Rossmeisl, John H; Davalos, R V (2010). "Non-thermal irreversible electroporation for deep intracranial disorders". 2010 Annual International Conference of the IEEE Engineering in Medicine and Biology: 2743–6. ISBN 978-1-4244-4123-5. PMID 21095962. doi:10.1109/IEMBS.2010.5626371

- Becker, S. M.; Kuznetsov, A. V. (2007). "Local Temperature Rises Influence in Vivo Electroporation Pore Development: A Numerical Stratum Corneum Lipid Phase Transition Model". Journal of Biomechanical Engineering. 129 (5): 712–21. PMID 17887897. doi:10.1115/1.2768380

- Regalado, Antonio (10 February 2017). "First Gene Drive in Mammals Could Aid Vast New Zealand Eradication Plan". MIT Tech Review. Retrieved 14 February 2017

- Weaver, James C.; Chizmadzhev, Yu.A. (1996). "Theory of electroporation: A review". Bioelectrochemistry and Bioenergetics. 41 (2): 135–60. doi:10.1016/S0302-4598(96)05062-3

Genetically Modified Organisms: An Overview

Genetically modified organisms are organisms whose genome has been altered. The modification consists of insertion, deletion and the mutation of genes. Genetically modified bacteria, genetically modified virus, genetically modified mammal, genetically modified insect and genetically modified fish are the types of organisms that humans have genetically modified. This section is an overview of the subject matter incorporating all the major aspects of genetically modified organisms.

Genetically Modified Organism

GloFish, the first genetically modified animal to be sold as a pet

A genetically modified organism (GMO) is any organism whose genetic material has been altered using genetic engineering techniques (i.e., a genetically *engineered* organism). GMOs are used to produce many medications and genetically modified foods and are widely used in scientific research and the production of other goods. The term GMO is very close to the technical legal term, 'living modified organism', defined in the Cartagena Protocol on Biosafety, which regulates international trade in living GMOs (specifically, "any living organism that possesses a novel combination of genetic material obtained through the use of modern biotechnology").

A more specifically defined type of GMO is a "transgenic organism." This is an organism whose genetic makeup has been altered by the addition of genetic material from an unrelated organism. This should not be confused with the more general way in which "GMO" is used to classify genetically altered organisms, as typically GMOs are organisms whose genetic makeup has been altered without the addition of genetic material from an unrelated organism.

The first genetically modified mouse was created in 1974, and the first plant was produced in 1983.

Production

Genetic modification involves the mutation, insertion, or deletion of genes. Inserted genes usually come from a different species in a form of horizontal gene-transfer. In nature this can occur when exogenous DNA penetrates the cell membrane for any reason. This can be accomplished artificially by:

- attaching the genes to a virus.

- physically inserting the extra DNA into the nucleus of the intended host with a very small syringe.

- using electroporation (that is, introducing DNA from one organism into the cell of another by use of an electric pulse).

- firing small particles from a gene gun.

Other methods exploit natural forms of gene transfer, such as the ability of *Agrobacterium* to transfer genetic material to plants, or the ability of lentiviruses to transfer genes to animal cells.

History

Herbert Boyer (pictured) and Stanley Cohen created the first genetically modified organism in 1973

Humans have domesticated plants and animals since around 12,000 BCE, using selective breeding or artificial selection (as contrasted with natural selection). The process of selective breeding, in which organisms with desired traits (and thus with the desired genes) are used to breed the next generation and organisms lacking the trait are not bred, is a precursor to the modern concept of genetic modification. Various advancements in genetics allowed humans to directly alter the DNA and therefore genes of organisms. In 1972 Paul Berg created the first recombinant DNA molecule when he combined DNA from a monkey virus with that of the lambda virus.

Herbert Boyer and Stanley Cohen made the first genetically modified organism (GMO) in 1973. They took a gene from a bacterium that provided resistance to the antibiotic kanamycin, inserted it into a plasmid and then induced another bacteria to uptake the plasmid. The bacteria was then able to survive in the presence of kanamycin. Boyer and Cohen expressed other genes in bacteria. This included genes from the toad Xenopus laevis in 1974, creating the first GMO expressing a gene from an organism from different kingdom.

In 1974 Rudolf Jaenisch created a transgenic mouse by introducing foreign DNA into its embryo, making it the world's first transgenic animal. However it took another eight years before transgenic mice were developed that passed the transgene to their offspring. Genetically modified mice were created in 1984 that carried cloned oncogenes, predisposed them to developing cancer. Mice with genes knocked out (knockout mouse) were created in 1989. The first transgenic livestock were produced in 1985 and the first animal to synthesise transgenic proteins in their milk were mice, engineered to produce human tissue plasminogen activator in 1987.

In 1983 the first genetically engineered plant was developed by Michael W. Bevan, Richard B. Flavell and Mary-Dell Chilton. They infected tobacco with *Agrobacterium* transformed with an antibiotic resistance gene and through tissue culture techniques were able to grow a new plant containing the resistance gene. The gene gun was invented in 1987, allowing transformation of plants not susceptible to *Agrobacterium* infection. In 2000, Vitamin A-enriched golden rice, was the first plant developed with increased nutrient value.

In 1976 Genentech, the first genetic engineering company was founded by Herbert Boyer and Robert Swanson; a year later, the company produced a human protein (somatostatin) in *E.coli*. Genentech announced the production of genetically engineered human insulin in 1978. The insulin produced by bacteria, branded humulin, was approved for release by the Food and Drug Administration in 1982. In 1988 the first human antibodies were produced in plants. In 1987, the ice-minus strain of *P. syringae* became the first genetically modified organism to be released into the environment when a strawberry field and a potato field in California were sprayed with it.

The first genetically modified crop, an antibiotic-resistant tobacco plant, was produced in 1982. China was the first country to commercialize transgenic plants, introducing a virus-resistant tobacco in 1992. In 1994 Calgene attained approval to commercially release the Flavr Savr tomato, the first genetically modified food. Also in 1994, the European Union approved tobacco engineered to be resistant to the herbicide bromoxynil, making it the first genetically engineered crop commercialized in Europe. An insect resistant Potato was approved for release in the USA in 1995, and by 1996 approval had been granted to commercially grow 8 transgenic crops and one flower crop (carnation) in 6 countries plus the EU.

In 2010, scientists at the J. Craig Venter Institute, announced that they had created the first synthetic bacterial genome. They named it Synthia and it was the world's first synthetic life form.

The first genetically modified animal to be commercialised was the GloFish, a Zebra fish with a fluorescent gene added that allows it to glow in the dark under ultraviolet light. The first genetically modified animal to be approved for food use was AquAdvantage salmon in 2015. The salmon were transformed with a growth hormone-regulating gene from a Pacific Chinook salmon and a promoter from an ocean pout enabling it to grow year-round instead of only during spring and summer.

Uses

GMOs are used in biological and medical research, production of pharmaceutical drugs, experimental medicine (e.g. gene therapy and vaccines against the Ebola virus), and agriculture (e.g. golden rice, resistance to herbicides), with developing uses in conservation. The term "genetically modified organism" does not always imply, but can include, targeted insertions of genes from one species into another. For example, a gene from a jellyfish, encoding a fluorescent protein called GFP, or green fluorescent protein, can be physically linked and thus co-expressed with mammalian genes to identify the location of the protein encoded by the GFP-tagged gene in the mammalian cell. Such methods are useful tools for biologists in many areas of research, including those who study the mechanisms of human and other diseases or fundamental biological processes in eukaryotic or prokaryotic cells.

Microbes

Bacteria

Bacteria were the first organisms to be modified in the laboratory, due to the relative ease of modifying their genetics.

They continue to be important model organisms for experiments in genetic engineering. In the field of synthetic biology, they have been used to test various synthetic approaches, from synthesizing genomes to creating novel nucleotides.

These organisms are now used for several purposes, and are particularly important in producing large amounts of pure human proteins for use in medicine.

Genetically modified bacteria are used to produce the protein insulin to treat diabetes. Similar bacteria have been used to produce biofuels, clotting factors to treat haemophilia, and human growth hormone to treat various forms of dwarfism.

Virus

In 2017 researchers genetically modified a virus to express spinach defensin proteins. The virus was injected into orange trees to combat citrus greening disease that had reduced orange production 70% since 2005.

Other

In addition, various genetically engineered micro-organisms are routinely used as sources of enzymes for the manufacture of a variety of processed foods. These include alpha-amylase from bacteria, which converts starch to simple sugars, chymosin from bacteria or fungi, which clots milk protein for cheese making, and pectinesterase from fungi, which improves fruit juice clarity.

Plants

Transgenic Plants

Transgenic plants have been engineered for scientific research, to create new colours in plants, and to create different crops.

In research, plants are engineered to help discover the functions of certain genes. One way to do this is to knock out the gene of interest and see what phenotype develops. Another strategy is to attach the gene to a strong promoter and see what happens when it is over expressed. A common technique used to find out where the gene is expressed is to attach it to GUS or a similar reporter gene that allows visualisation of the location.

Kenyans examining insect-resistant transgenic Bt corn

After thirteen years of collaborative research, an Australian company – Florigene, and a Japanese company – Suntory, created a blue rose (actually lavender or mauve) in 2004. The genetic engineering involved three alterations – adding two genes, and interfering with another. One of the added genes was for the blue plant pigment delphinidin cloned from the pansy. The researchers then used RNA interference (RNAi) technology to depress all color production by endogenous genes by blocking a crucial protein in color production, called dihydroflavonol 4-reductase) (DFR), and adding a variant of that protein that would not be blocked by the RNAi but that would allow the delphinidin to work. The roses are sold in Japan, the United States, and Canada. Florigene has also created and sells lavender-colored carnations that are genetically engineered in a similar way.

Suntory "blue" rose

Simple plants and plant cells have been genetically engineered for production of biopharmaceuticals in bioreactors as opposed to cultivating plants in open fields. Work has been done with duckweed *Lemna minor*, the algae *Chlamydomonas reinhardtii* and the moss *Physcomitrella patens*. An Israeli company, Protalix, has developed a method to produce therapeutics in cultured transgenic carrot and tobacco cells. Protalix and its partner, Pfizer, received FDA approval to market its drug Elelyso, a treatment for Gaucher's disease, in 2012.

Genetically Modified Crops

Genetically modified crops (GM crops, or biotech crops) are plants used in agriculture, the DNA of which has been modified using genetic engineering techniques. In most cases the aim is to introduce a new trait to the plant which does not occur naturally in the species. Examples in food crops include resistance to certain pests, diseases, or environmental conditions, reduction of spoilage, or resistance to chemical treatments (e.g. resistance to a herbicide), or improving the nutrient profile of the crop. Examples in non-food crops include production of pharmaceutical agents, biofuels, and other industrially useful goods, as well as for bioremediation.

Farmers have widely adopted GM technology. Between 1996 and 2013, the total surface area of land cultivated with GM crops increased by a factor of 100, from 17,000 square kilometers (4,200,000 acres) to 1,750,000 km² (432 million acres). 10% of the world's croplands were planted with GM crops in 2010. In the US, by 2014, 94% of the planted area of soybeans, 96% of cotton and 93% of corn were genetically modified varieties. In recent years GM crops expanded rapidly in developing countries. In 2013 approximately 18 million farmers grew 54% of worldwide GM crops in developing countries.

Cisgenic Plants

Cisgenesis, sometimes also called intragenesis, is a product designation for a category of genetically engineered plants. A variety of classification schemes have been proposed that order genetically modified organisms based on the nature of introduced genotypical changes rather than the process of genetic engineering.

While some genetically modified plants are developed by the introduction of a gene originating from distant, sexually incompatible species into the host genome, cisgenic plants contain genes that have been isolated either directly from the host species or from sexually compatible species. The new genes are introduced using recombinant DNA methods and gene transfer. Some scientists hope that the approval process of cisgenic plants might be simpler than that of proper transgenics, but it remains to be seen.

Conservation in Plants

Genetically modified organisms have been proposed to aid conservation of plant species threatened by extinction. Many trees face the threat of invasive plants and diseases, such as the emerald ash borer in North American and the fungal disease, Ceratocystis platani, in European plane trees. A suggested solution to increase the resilience of threatened tree species is to genetically modify individuals by transferring resistant genes. Papaya trees are an example of a species that was successfully conserved using genetic modification. The papaya ringspot virus (PRSV) devastated papaya trees in Hawaii in the twentieth century until transgenic papaya plants were given pathogen-derived resistance.

However, genetic modification for conservation in plants remains mainly speculative and further experimentation is needed before the technique can be widely implemented. A main concern with using genetic modification for conservation purposes is that a transgenic species may no longer bear enough resemblance to the original species to truly claim that the original species is being conserved. Instead, the transgenic species may be genetically different enough to be considered a new species, thus diminishing the conservation worth of genetic modification.

Mammals

Genetically modified mammals are an important category of genetically modified organisms. Ralph L. Brinster and Richard Palmiter developed the techniques responsible for transgenic mice, rats, rabbits, sheep, and pigs in the early 1980s, and established many of the first transgenic models of human disease, including the first carcinoma caused by a transgene. The process of genetically engineering animals is a slow, tedious, and expensive process. However, new technologies are making genetic modifications easier and more precise.

The first transgenic (genetically modified) animal was produced by injecting DNA into mouse embryos then implanting the embryos in female mice.

Genetically modified animals currently being developed can be placed into six different broad classes based on the intended purpose of the genetic modification:

1. to research human diseases (for example, to develop animal models for these diseases);

2. to produce industrial or consumer products (fibres for multiple uses);

3. to produce products intended for human therapeutic use (pharmaceutical products or tissue for implantation);

4. to enrich or enhance the animals' interactions with humans (hypo-allergenic pets);

5. to enhance production or food quality traits (faster growing fish, pigs that digest food more efficiently);

6. to improve animal health (disease resistance)

Research Use

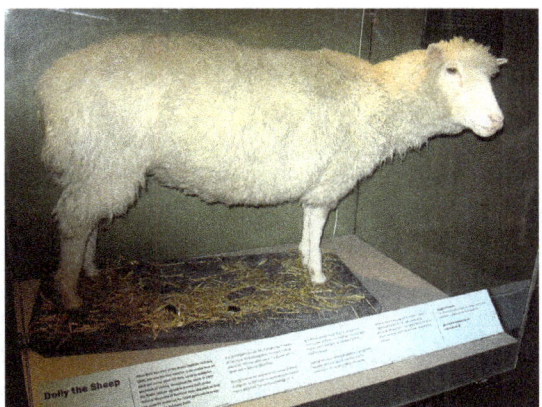

Dolly was a female domestic sheep and the first animal to be cloned from an adult somatic cell

Transgenic animals are used as experimental models to perform phenotypic and for testing in biomedical research.

Genetically modified (genetically engineered) animals are becoming more vital to the discovery and development of cures and treatments for many serious diseases. By altering the DNA or transferring DNA to an animal, we can develop certain proteins that may be used in medical treatment. Stable expressions of human proteins have been developed in many animals, including sheep, pigs, and rats. Human-alpha-1-antitrypsin, which has been tested in sheep and is used in treating humans with this deficiency and transgenic pigs with human-histo-compatibility have been studied in the hopes that the organs will be suitable for transplant with less chances of rejection.

Scientists have genetically engineered several organisms, including some mammals, to include green fluorescent protein (GFP), first observed in the jellyfish, *Aequorea victoria* in 1962, for medical research purposes (Chalfie, Shimoura, and Tsien were awarded the Nobel prize in Chemistry in 2008 for the discovery and development of GFP). For example, fluorescent pigs have been bred to study human organ transplants (xenotransplantation), regenerating ocular photoreceptor cells, and other topics. In 2011 a Japanese-American team created green-fluorescent cats to find therapies for HIV/AIDS and other diseases as feline immunodeficiency virus (FIV) is related to HIV.

In 2009, scientists in Japan announced that they had successfully transferred a gene into a primate species (marmosets) and produced a stable line of breeding transgenic primates for the first time. Their first research target for these marmosets was Parkinson's disease, but they were also considering amyotrophic lateral sclerosis and Huntington's disease.

Producing Human Therapeutics

Within the field known as pharming, intensive research has been conducted to develop transgenic animals that produce biotherapeutics. On 6 February 2009, the U.S. Food and Drug Administration approved the first human biological drug produced from such an animal, a goat. The drug, ATryn, is an anticoagulant which reduces the probability of blood clots during surgery or childbirth. It is extracted from the goat's milk.

Production or Food Quality Traits

In 2006, a pig was engineered to produce omega-3 fatty acids through the expression of a roundworm gene.

Enviropig was a genetically enhanced line of Yorkshire pigs in Canada created with the capability of digesting plant phosphorus more efficiently than conventional Yorkshire pigs. The project ended in 2012. These pigs produced the enzyme phytase, which breaks down the indigestible phosphorus, in their saliva. The enzyme was introduced into the pig chromosome by pronuclear microinjection. With this enzyme, the animal is able to digest cereal grain phosphorus. The use of these pigs would reduce the potential of water pollution since they excrete from 30 to 70.7% less phosphorus in manure depending upon the age and diet. The lower concentrations of phosphorus in surface runoff reduces algal growth, because phosphorus is the limiting nutrient for algae. Because algae consume large amounts of oxygen, it can result in dead zones for fish.

In 2011, Chinese scientists generated dairy cows genetically engineered with genes from human beings to produce milk that would be the same as human breast milk. This could potentially benefit mothers who cannot produce breast milk but want their children to have breast milk rather than formula. Aside from milk production, the researchers claim these transgenic cows to be identical to regular cows. Two months later scientists from Argentina presented Rosita, a transgenic cow incorporating two human genes, to produce milk with similar properties as human breast milk. In 2012, researchers from New Zealand also developed a genetically engineered cow that produced allergy-free milk.

Goats have been genetically engineered to produce milk with strong spiderweb-like silk proteins in their milk.

Human Gene Therapy

Gene therapy, uses genetically modified viruses to deliver genes that can cure disease in humans. Although gene therapy is still relatively new, it has had some successes. It has been used to treat genetic disorders such as severe combined immunodeficiency, and Leber's congenital amaurosis. Treatments are also being developed for a range of other currently incurable diseases, such as cystic fibrosis, sickle cell anemia, Parkinson's disease, cancer, diabetes, heart disease and muscular dystrophy.

Conservation Use

Genetically modified organisms have been used to conserve European wild rabbits in the Iberian peninsula and Australia. In both cases, the genetically modified organism used was a myxoma virus, but for opposite purposes: to protect the endangered population in Europe with immunizations and to regulate the overabundant population in Australia with contraceptives.

In the Iberian peninsula, the European wild rabbit population has experienced a sharp decline from viral diseases and overhunting. To protect the species from viral diseases, the myxoma virus was genetically modified to immunize the rabbits. The European wild rabbit population in Australia faces the opposite problem: lack of natural predators has made the introduced species invasive. The same myxoma virus was genetically modified to lower fertility in the Australian rabbit population.

Fish

Genetically modified fish are used for scientific research and as pets, and are being considered for use as food and as aquatic pollution sensors.

GM fish are widely used in basic research in genetics and development. Two species of fish, zebrafish and medaka, are most commonly modified because they have optically clear chorions (membranes in the egg), rapidly develop, and the 1-cell embryo is easy to see and microinject with transgenic DNA.

The GloFish is a patented brand of genetically modified (GM) fluorescent zebrafish with bright red, green, and orange fluorescent color. Although not originally developed for the ornamental fish trade, it became the first genetically modified animal to become publicly available as a pet when it was introduced for sale in 2003. They were quickly banned for sale in California.

GM fish have been developed with promoters driving an over-production of "all fish" growth hormone for use in the aquaculture industry to increase the speed of development and potentially reduce fishing pressure on wild stocks. This has resulted in dramatic growth enhancement in several species, including salmon, trout and tilapia. AquaBounty Technologies, a biotechnology company working on bringing a GM salmon to market, claims that their GM AquAdvantage salmon can mature in half the time as wild salmon. AquaBounty applied for regulatory approval to market their GM salmon in the US, and was approved in November 2015. On 25 November 2013 Canada approved commercial scale production and export of GM Salmon eggs but they are not approved for human consumption in Canada.

Several academic groups have been developing GM zebrafish to detect aquatic pollution. The lab that originated the GloFish discussed above originally developed them to change color in the presence of pollutants, to be used as environmental sensors. A lab at University of Cincinnati has been developing GM zebrafish for the same purpose, as has a lab at Tulane University.

Recent research on pain in fish has resulted in concerns being raised that genetic-modifications induced for scientific research may have detrimental effects on the welfare of fish.

Frogs

Genetically modified frogs are used for scientific research and are widely used in basic research including genetics and early development. Two species of frog, *Xenopus laevis* and *Xenopus tropicalis*, are most commonly used.

GM frogs are also being used as pollution sensors, especially for endocrine disrupting chemicals.

Invertebrates

Fruit Flies

In biological research, transgenic fruit flies (*Drosophila melanogaster*) are model organisms used to study the effects of genetic changes on development. Fruit flies are often preferred over other animals due to their short life cycle, low maintenance requirements, and relatively simple genome compared to many vertebrates.

Mosquitoes

In 2010, scientists created "malaria-resistant mosquitoes" in the laboratory. The World Health Organization estimated that malaria killed almost one million people in 2008. Genetically modified male mosquitoes containing a lethal gene have been developed to combat the spread of dengue fever and the Zika virus. *Aedes aegypti* mosquitoes, the single most important carrier of dengue fever and the Zika virus, were reduced by 80% in a 2010 trial of these GM mosquitoes in the Cayman Islands and by 90% in a 2015 trial in Bahia, Brazil. In comparison, the Florida Keys Mosquito Control District has achieved only 30%-60% population reduction with traps and pesticide spraying. In 2016 FDA approved a genetically modified mosquito intervention for Key West, Florida. UK firm Oxitec proposed the release of millions of modified male (non-biting) mosquitoes to compete with wild males for mates. The males are engineered so that their offspring die before maturing, helping to eradicate mosquito-borne disease. Final approval was to be based on a local referendum to be held in November. Andrea Crisanti, a molecular biologist at Imperial College in London is working on ways to stop the A. gambiae mosquito from transmitting disease.

Bollworms

A strain of *Pectinophora gossypiella* (Pink bollworm) has been genetically engineered to express a red fluorescent protein. This allows researchers to monitor bollworms that have been sterilized by radiation and released to reduce bollworm infestation. The strain has been field tested for over three years and has been approved for release.

Cnidaria

Cnidaria such as *Hydra* and the sea anemone *Nematostella vectensis* are attractive model organisms to study the evolution of immunity and certain developmental processes. An important technical breakthrough was the development of procedures for generation of stable transgenic hydras and sea anemones by embryo microinjection.

Regulation

The regulation of genetic engineering concerns the approaches taken by governments to assess and manage the risks associated with the use of genetic engineering technology and the development and release of genetically modified organisms (GMO), including genetically modified crops and genetically modified fish. There are differences in the regulation of GMOs between countries, with some of the most marked differences occurring between the USA and Europe. Regulation varies in a given country depending on the intended use of the products of the genetic engineering. For example, a crop not intended for food use is generally not reviewed by authorities responsible for food safety. The European Union differentiates between approval for cultivation within the EU and approval for import and processing. While only a few GMOs have been approved for cultivation in the EU a number of GMOs have been approved for import and processing. The cultivation of GMOs has triggered a debate about the market for GMOs in Europe. Depending on the coexistence regulations, incentives for cultivation of GM crops differ.

Controversy

There is controversy over GMOs, especially with regard to their use in producing food. The dispute involves buyers, biotechnology companies, governmental regulators, nongovernmental organizations, and scientists. The key areas of controversy related to GMO food are whether GM food should be labeled, the role of government regulators, the effect of GM crops on health and the environment, the effect on pesticide resistance, the impact of GM crops for farmers, and the role of GM crops in feeding the world population. In 2014, sales of products that had been labeled as non-GMO grew 30 percent to $1.1 billion.

There is a scientific consensus that currently available food derived from GM crops poses no greater risk to human health than conventional food, but that each GM food needs to be tested on a case-by-case basis before introduction. Nonetheless, members of the public are much less likely than scientists to perceive GM foods as safe. The legal and regulatory status of GM foods varies by country, with some nations banning or restricting them, and others permitting them with widely differing degrees of regulation.

No reports of ill effects have been proven in the human population from ingesting GM food. Although labeling of GMO products in the marketplace is required in many countries, it is not required in the United States and no distinction between marketed GMO and non-GMO foods is recognized by the US FDA. In a May 2014 article in *The Economist* it was argued that, while GM foods could potentially help feed 842 million malnourished people globally, laws such as the one passed in Vermont, to require labeling of foods containing genetically modified ingredients, could have the unintended consequence of interrupting the process of spreading GM technologies to impoverished countries that suffer with food security problems.

A 2014 critical review of histopathology studies on rats (eating approved widely eaten GM crops) found significant flaws, inadequacies, and a lack of transparency in methodology and results. Published studies could be found for only 19% of these widely eaten crops. Most of reviewed studies were performed after the approval of crop. More research is needed, including long-term animal feeding studies and thorough histopathological investigations.

The Organic Consumers Association, and the Union of Concerned Scientists, and Greenpeace stated that risks have not been adequately identified and managed, and they have questioned the objectivity of regulatory authorities. Some health groups say there are unanswered questions regarding the potential long-term impact on human health from food derived from GMOs, and propose mandatory labeling or a moratorium on such products. Concerns include contamination of the non-genetically modified food supply, effects of GMOs on the environment and nature, the rigor of the regulatory process, and consolidation of control of the food supply in companies that make and sell GMOs, or concerns over the use of herbicides with glyphosate.

Genetically Modified Bacteria

Genetically modified bacteria were the first organisms to be modified in the laboratory, due to their simple genetics. These organisms are now used for several purposes, and are particularly important in producing large amounts of pure human proteins for use in medicine.

History

The first example of this occurred in 1978 when Herbert Boyer, working at a University of California laboratory, took a version of the human insulin gene and inserted into the bacterium *Escherichia coli* to produce synthetic "human" insulin. Four years later, it was approved by the U.S. Food and Drug Administration.

Pharmaceutical Production

The drug industry has made use of this discovery to produce medication for diabetes. Similar bacteria have been used to produce clotting factors to treat haemophilia, and human growth hormone to treat various forms of dwarfism. These recombinant proteins are safer than the products they replaced. Prior to recombinant protein products, several treatments were derived from cadavers or other donated body fluids and could transmit diseases. Indeed, transfusion of blood products had previously led to unintentional infection of haemophiliacs with HIV or hepatitis C; similarly, treatment with human growth hormone derived from cadaver pituitary glands may have led to outbreaks of Creutzfeldt–Jakob disease.

Other Uses

Genetically modified bacteria can serve various purposes beyond producing medicinal compounds. For instance, bacteria which generally cause tooth decay have been engineered to no longer produce tooth-corroding lactic acid. These transgenic bacteria, if allowed to colonize a person's mouth, could perhaps reduce the formation of cavities. Transgenic microbes have also been used in recent research to kill or hinder tumors, and to fight Crohn's disease. Genetically modified bacteria are also used in some soils to facilitate crop growth, and can also produce chemicals toxic to crop pests.

GM bacteria have also been developed to leach copper from ore, clean up mercury pollution and detect arsenic in drinking water.

Bacteria-synthesized Transgenic Products

- Insulin
- Hepatitis B vaccine
- Tissue plasminogen activator
- Human growth hormone
- Ice-minus bacteria
- Interferon
- Bt corn
- Terraforming

Genetically Modified Virus

A genetically modified virus is a virus that has gone through genetic modification for various biomedical purposes.

General Usage

Genetic modification involves the insertion or deletion of genes. When genes are inserted, they usually come from different species, which is a form of horizontal gene transfer. In nature this can occur when exogenous DNA penetrates the cell membrane for any reason but usually for domination of other diseases.

To do this artificially may require attaching the genes to a virus or just physically inserting the extra DNA into the nucleus of the intended host with a very small syringe, or with very small particles fired from a gene gun. However, other methods exploit natural forms of gene transfer, such as the ability of *Agrobacterium* to transfer genetic material to plants, or the ability of lentiviruses to transfer genes to animal cells.

Lithium-ion Batteries

In materials science, a genetically modified virus has been used to construct a more environmentally friendly lithium-ion battery.

Gene Therapy

Gene therapy uses genetically modified viruses to deliver genes that can cure disease into human cells. Although gene therapy is still relatively new, it has had some successes. It has been used to treat genetic disorders such as severe combined immunodeficiency.

Heart Pacemaker

In 2012, US researchers reported that they injected a genetically modified virus into the heart of guinea pigs. This virus inserted into the heart muscles a gene called Tbx18 which enabled heartbeats. The researchers forecast that one day this technique could be used to restore the heartbeat in humans who would otherwise need electronic pacemakers.

Cancer Treatment

In 2004, researchers reported that a genetically modified virus that exploits the selfish behaviour of cancer cells might offer an alternative way of killing tumours. Since then, several researchers have developed genetically modified oncolytic viruses that show promise as treatments for various types of cancer.

Rabbits

In Spain and Portugal, by 2005 rabbits had declined by as much as 95% over 50 years due diseases such as myxomatosis, rabbit haemorrhagic disease and other causes. This in turn caused declines in predators like the Iberian lynx, a critically endangered species. In 2000 Spanish researchers investigated a genetically modified virus which might have protected rabbits in the wild against myxomatosis and rabbit haemorrhagic disease. However, there was concern that such a virus might make its way into wild populations in areas such as Australia and create a population boom. Rabbits in Australia are considered to be such a pest that land owners are legally obliged to control them.

GMO lentivirus

A scientist claims she was infected by a genetically modified virus while working for Pfizer. In her federal lawsuit she says she has been intermittently paralyzed by the Pfizer-designed virus. "McClain, of Deep River, suspects she was inadvertently exposed, through work by a former Pfizer colleague in 2002 or 2003, to an engineered form of the lentivirus, a virus similar to the one that can lead to acquired immune deficiency syndrome, or AIDS." The court found that McClain failed to demonstrate that her illness was caused by exposure to the lentivirus, but also that Pfizer violated whistleblower laws.

Biohazard Research Limitations

The National Institute of Health declared a research funding moratorium on select Gain-of-Function virus research in January 2015. Questions about a potential escape of a modified virus from a biosafety lab and the utility of dual-use-technology, dual use research of concern (DURC), prompted the NIH funding policy revision.

Genetically Modified Mammal

Genetically modified mammals are mammals that have been genetically engineered. They are an important category of genetically modified organisms. The majority of research involving genetically modified mammals involves mice with attempts to produce knockout animals in other mammalian species limited by the inability to derive and stably culture embryonic stem cells.

Usage

The majority of genetically modified mammals are used in research to investigate changes in phenotype when specific genes are altered. This can be used to discover the function of an unknown gene, any genetic interactions that occur or where the gene is expressed. Genetic modification can also produce mammals that are susceptible to certain compounds or stresses for testing in biomedical research. Some genetically modified mammals are used as models of human diseases and potential treatments and cures can first be tested on them. Other mammals have been engineered with the aim of potentially increasing their use to medicine and industry. These possibilities include pigs expressing human antigens aiming to increasing the success of xenotransplantation to lactating mammals expressing useful proteins in their milk.

Genetically Modified Mice

Genetically modified mice are often used to study cellular and tissue-specific responses to disease (cf knockout mouse). This is possible since mice can be created with the same mutations that occur in human genetic disorders, the production of the human disease in these mice then allows treatments to be tested.

The oncomouse is a type of laboratory mouse that has been genetically modified developed by Philip Leder and Timothy A. Stewart of Harvard University to carry a specific gene called an activated oncogene.

Metabolic supermice are the creation of a team of American scientists led by Richard Hanson, professor of biochemistry at Case Western Reserve University at Cleveland, Ohio. The aim of the research was to gain a greater understanding of the PEPCK-C enzyme, which is present mainly in the liver and kidneys.

Genetically Modified Rats

A knockout rat is a rat with a single gene disruption used for academic and pharmaceutical research.

Genetically modified Goats

BioSteel is a trademark name for a high-strength based fiber material made of the recombinant spider silk-like protein extracted from the milk of transgenic goats, made by Nexia Biotechnologies. The company has successfully generated distinct lines of goats that produce in their milk recombinant versions of either the MaSpI or MaSpII dragline silk proteins, respectively. Nexia Biotechnologies, however, went bankrupt and is no longer company.

Genetically Modified Pigs

The enviropig is the trademark for a genetically modified line of Yorkshire pigs with the capability to digest plant phosphorus more efficiently than ordinary unmodified pigs that was developed at the University of Guelph. Enviropigs produce the enzyme phytase in the salivary glands that is secreted in the saliva.

In 2006 the scientists from National Taiwan University's Department of Animal Science and Technology managed to breed three green-glowing pigs using green fluorescent protein. Fluorescent pigs can be used to study human organ transplants, regenerating ocular photoreceptor cells, neuronal cells in the brain, regenerative medicine via stem cells, tissue engineering, and other diseases.

In 2015, researchers at the Beijing Genomics Institute used transcription activator-like effector nucleases to create a miniature version of the Bama breed of pigs, and offered them for sale to consumers.

In 2017 scientists at the Roslin Institute of the University of Edinburgh, reported they had bred pigs with a modified CD163 gene. These pigs were completely resistant to Porcine Reproductive and Respiratory Syndrome, a disease that causes major losses in the world-wide pig industry.

Genetically Modified Cattle

Herman the Bull was in 1991 the first genetically modified or transgenic bovine in the world. The announcement of Herman's creation caused an ethical storm.

Genetically Modified Dogs

Ruppy (short for Ruby Puppy) was in 2009 the world's first Genetically modified dog. A cloned beagle, Ruppy and four other beagles produced a fluorescent protein that glowed red upon excitation with ultraviolet light. It was hoped to use this procedure to investigate the effect of the hormone oestrogen on fertility.

A team in China reported in 2015 that they had genetically engineered beagles to have twice the normal muscle mass, inserting a natural myostatin gene mutation taken from whippets.

Genetically Modified Primates

In 2009 scientists in Japan announced that they had successfully transferred a gene into a primate species (marmosets) and produced a stable line of breeding transgenic primates for the first time. It was hoped that this would aid research into human diseases that cannot be studied in mice, for example Huntington's disease, strokes, Alzheimer's disease and schizophrenia.

Genetically Modified Cats

In 2011 a Japanese-American Team created genetically modified green-fluorescent cats in order to find therapies for HIV/AIDS and other diseases as Feline immunodeficiency virus (FIV) is related to HIV.

Knockout Mouse

A knockout mouse or knock-out mouse is a genetically modified mouse (*Mus musculus*) in which researchers have inactivated, or "knocked out", an existing gene by replacing it or disrupting it with an artificial piece of DNA. They are important animal models for studying the role of genes which have been sequenced but whose functions have not been determined. By causing a specific gene to be inactive in the mouse, and observing any differences from normal behaviour or physiology, researchers can infer its probable function.

Mice are currently the laboratory animal species most closely related to humans for which the knockout technique can easily be applied. They are widely used in knockout experiments, especially those investigating genetic questions that relate to human physiology. Gene knockout in rats is much harder and has only been possible since 2003.

The first recorded knockout mouse was created by Mario R. Capecchi, Martin Evans, and Oliver Smithies in 1989, for which they were awarded the 2007 Nobel Prize in Physiology or Medicine. Aspects of the technology for generating knockout mice, and the mice themselves have been patented in many countries by private companies.

Use

Knocking out the activity of a gene provides information about what that gene normally does. Humans share many genes with mice. Consequently, observing the characteristics of knockout mice gives researchers information that can be used to better understand how a similar gene may cause or contribute to disease in humans.

Examples of research in which knockout mice have been useful include studying and modeling different kinds of cancer, obesity, heart disease, diabetes, arthritis, substance abuse, anxiety, aging and Parkinson's disease. Knockout mice also offer a biological and scientific context in which drugs and other therapies can be developed and tested.

Millions of knockout mice are used in experiments each year.

Strains

There are several thousand different strains of knockout mice. Many mouse models are named after the gene that has been inactivated. For example, the p53 knockout mouse is named after the p53 gene which codes for a protein that normally suppresses the growth of tumours by arresting cell division and/or inducing apoptosis. Humans born with mutations that deactivate the p53 gene suffer from Li-Fraumeni syndrome, a condition that dramatically increases the risk of developing bone cancers, breast cancer and blood cancers at an early age. Other mouse models are named according to their physical characteristics or behaviours.

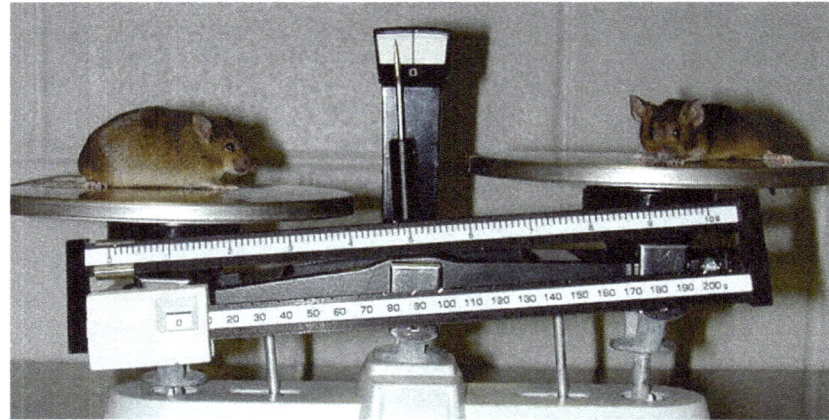

A knockout mouse (left) that is a model for obesity, compared with a normal mouse.

Procedure

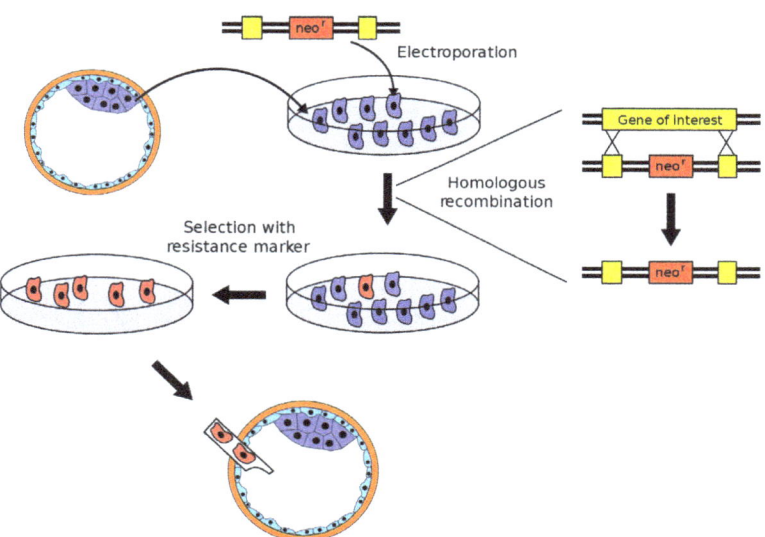

The procedure for making mixed-genotype blastocyst.

There are several variations to the procedure of producing knockout mice; the following is a typical example:

1. The gene to be knocked out is isolated from a mouse gene library. Then a new DNA sequence is engineered which is very similar to the original gene and its immediate neighbour sequence, except that it is changed sufficiently to make the gene inoperable. Usually, the new sequence is also given a marker gene, a gene that normal mice don't have and that

confers resistance to a certain toxic agent (e.g., neomycin) or that produces an observable change (e.g. colour or fluorescence). In addition, a second gene, such as herpes tk+, is also included in the construct in order to accomplish a complete selection.

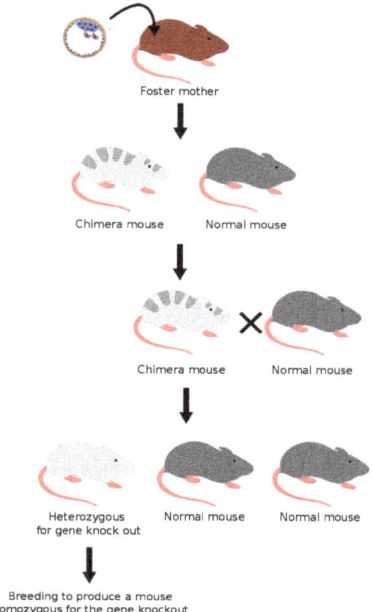

Breeding scheme for producing knockout mice. Blastocysts containing cells, that are both wildtype and knockout cells, are injected into the uterus of a foster mother. This produces offspring that are either wildtype and coloured the same colour as the blastocyst donor (grey) or chimera (mixed) and partially knocked out. The chimera mice are crossed with a normal wildtype mouse (grey). This produces offspring that are either white and heterozygous for the knocked out gene or grey and wildtype. White heterozygous mice can subsequently be crossed to produce mice that are homozygous for the knocked out gene.

2. Embryonic stem cells are isolated from a mouse blastocyst (a very young embryo) and grown *in vitro*. For this example, we will take stem cells from a white mouse.

3. The new sequence from step 1 is introduced into the stem cells from step 2 by electroporation. By the natural process of homologous recombination some of the electroporated stem cells will incorporate the new sequence with the knocked-out gene into their chromosomes in place of the original gene. The chances of a successful recombination event are relatively low, so the majority of altered cells will have the new sequence in only one of the two relevant chromosomes – they are said to be heterozygous. Cells that were transformed with a vector containing the neomycin resistance gene and the herpes tk+ gene are grown in a solution containing neomycin and Ganciclovir in order to select for the transformations that occurred via homologous recombination. Any insertion of DNA that occurred via random insertion will die because they test positive for both the neomycin resistance gene and the herpes tk+ gene, whose gene product reacts with Ganciclovir to produce a deadly toxin. Moreover, cells that do not integrate any of the genetic material test negative for both genes and therefore die as a result of poisoning with neomycin.

4. The embryonic stem cells that incorporated the knocked-out gene are isolated from the unaltered cells using the marker gene from step 1. For example, the unaltered cells can be killed using a toxic agent to which the altered cells are resistant.

5. The knocked-out embryonic stem cells from step 4 are inserted into a mouse blastocyst. For this example, we use blastocysts from a grey mouse. The blastocysts now contain two types of stem cells: the original ones (from the grey mouse), and the knocked-out cells (from the white mouse). These blastocysts are then implanted into the uterus of female mice, where they develop. The newborn mice will therefore be chimeras: some parts of their bodies result from the original stem cells, other parts from the knocked-out stem cells. Their fur will show patches of white and grey, with white patches derived from the knocked-out stem cells and grey patches from the recipient blastocyst.

6. Some of the newborn chimera mice will have gonads derived from knocked-out stem cells, and will therefore produce eggs or sperm containing the knocked-out gene. When these chimera mice are crossbred with others of the wild type, some of their offspring will have one copy of the knocked-out gene in all their cells. These mice will be entirely white and are not chimeras, however they are still heterozygous.

7. When these heterozygous offspring are interbred, some of their offspring will inherit the knocked-out gene from both parents; they carry no functional copy of the original unaltered gene (i.e. they are homozygous for that allele).

A detailed explanation of how knockout (KO) mice are created is located at the website of the Nobel Prize in Physiology or Medicine 2007.

Limitations

The National Institutes of Health discusses some important limitations of this technique.

While knockout mouse technology represents a valuable research tool, some important limitations exist. About 15 percent of gene knockouts are developmentally lethal, which means that the genetically altered embryos cannot grow into adult mice. This problem is often overcome through the use of conditional mutations. The lack of adult mice limits studies to embryonic development and often makes it more difficult to determine a gene's function in relation to human health. In some instances, the gene may serve a different function in adults than in developing embryos.

Knocking out a gene also may fail to produce an observable change in a mouse or may even produce different characteristics from those observed in humans in which the same gene is inactivated. For example, mutations in the p53 gene are associated with more than half of human cancers and often lead to tumours in a particular set of tissues. However, when the p53 gene is knocked out in mice, the animals develop tumours in a different array of tissues.

There is variability in the whole procedure depending largely on the strain from which the stem cells have been derived. Generally cells derived from strain 129 are used. This specific strain is not suitable for many experiments (e.g., behavioural), so it is very common to backcross the offspring to other strains. Some genomic loci have been proven very difficult to knock out. Reasons might be the presence of repetitive sequences, extensive DNA methylation, or heterochromatin. The confounding presence of neighbouring 129 genes on the knockout segment of genetic material has been dubbed the "flanking-gene effect". Methods and guidelines to deal with this problem have been proposed.

Another limitation is that conventional (i.e. non-conditional) knockout mice develop in the absence of the gene being investigated. At times, loss of activity during development may mask the role of the gene in the adult state, especially if the gene is involved in numerous processes spanning development. Conditional/inducible mutation approaches are then required that first allow the mouse to develop and mature normally prior to ablation of the gene of interest.

Another serious limitation is a lack of evolutive adaptations in knockout model that might occur in wild type animals after they naturally mutate. For instance, erythrocyte-specific coexpression of GLUT1 with stomatin constitutes a compensatory mechanism in mammals that are unable to synthesize vitamin C.

Knockout Rat

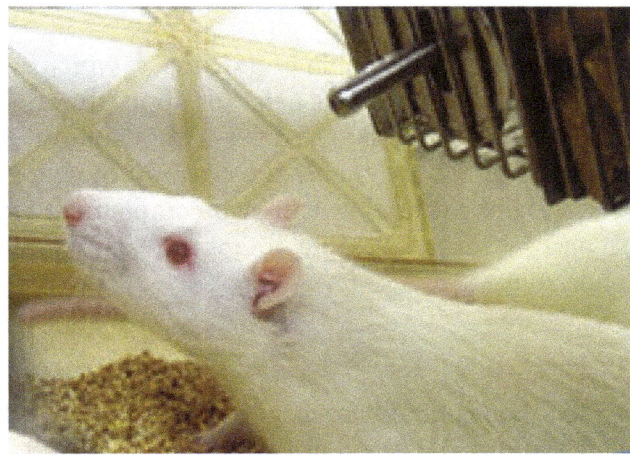

Knockout rat

A knockout rat is a genetically engineered rat with a single gene turned off through a targeted mutation (gene trapping) used for academic and pharmaceutical research. Knockout rats can mimic human diseases and are important tools for studying gene function (functional genomics) and for drug discovery and development. The production of knockout rats was not economically or technically feasible until 2008.

Technology developed through funding from the National Institutes of Health (NIH) and work accomplished by the members of the Knock Out Rat Consortium (KORC) led to cost-effective methods to create knockout rats. The importance of developing the rat as a more versatile tool for human health research is evidenced by the $120 million investment made by the NIH via the Rat Genome Sequencing Project Consortium, resulting in the draft sequence of a laboratory strain of the brown or Norway rat (*Rattus norvegicus*). Additional developments with zinc finger nuclease technology in 2009 led to the first knockout rat with targeted, germline-transmitted mutations. Knockout rat disease models for Parkinson's, Alzheimer's, hypertension, and diabetes using zinc-finger nuclease technology are being commercialized by SAGE Labs.

Research Use

Mice, rats, and humans share all but approximately 1% of each other's genes making rodents good model organisms for studying human gene function. Both mice and rats are relatively small, easily handled, have a short generation time, and are genetically inbred. While mice have proven to be a useful rodent model and techniques have been developed for routine disruption of their genes, in many circumstances rats are considered a superior laboratory animal for studying and modeling human disease.

Rats are physiologically more similar to humans than are mice. For example, rats have a heart rate more similar to that of humans, while mice have a heart rate five to ten times as fast. It is widely believed that the rat is a better model than the mouse for human cardiovascular disease, diabetes, arthritis, and many autoimmune, neurological, behavioral, and addiction disorders. In addition, rat models are superior to mouse models for testing the pharmacodynamics and toxicity of potential therapeutic compounds, partially because the number and type of many of their detoxifying enzymes are very similar to those in humans. Their larger size makes rats more conducive to study by instrumentation, and also facilitates manipulation such as blood sampling, nerve conduction, and performing surgeries.

Techniques for genetic manipulation are available in the mouse, which is commonly used to model human disease. Although published knockouts exist for approximately 60% of mouse genes, a large majority of common human diseases do not have a knockout mouse model. Knockout rat models are an alternative to mice that may enable the creation of new gene disruptions that are unavailable in the mouse. Knockout rat models can also complement existing transgenic mouse models. Comparing mouse and rat mutants can facilitate the distinction between rodent-specific and general mammalian phenotypes.

Production Challenges

Rat models have been used to advance many areas of medical research, including cardiovascular disease, psychiatric disorders (studies of behavioral intervention and addiction), neural regeneration, diabetes, transplantation, autoimmune disorders (rheumatoid arthritis), cancer, and wound & bone healing. While the completion of the rat genome sequence provides very key information, how these diseases relate to gene function requires an efficient method to create knockout rat models in which specific genomic sequences are manipulated. Most techniques for genetic manipulation, including random mutagenesis with a gene trap (retroviral-based and non-retroviral-based), gene knock-outs/knock-ins, and conditional mutations, depend upon the culture and manipulation of embryonic stem (ES) cells. Rat ES cells were only recently isolated and no demonstration of gene modification in them has been reported. Consequently, many genetic manipulation techniques widely used in the mouse are not possible in the rat.

Early Methods

Until the commercial development of mobile DNA technology in 2007 and zinc-finger nuclease technology in 2009, there were only two technologies that could be used to produce rat models of human disease: cloning and chemical mutagenesis using N-ethyl-N-nitrosourea (ENU). Although cloning by somatic cell nuclear transfer (SCNT) could theoretically be used to create rats with specific mutations by mutating somatic cells, and then using these cells for SCNT, this approach has not been used successfully to create knockout rats. One problem with this strategy is that SCNT is extremely inefficient. The first published attempt had a success rate of less than 1%. Alternatively, ENU mutagenesis is a common random mutagenesis gene knockout strategy in the mouse that can also be used in the rat. ENU mutagenesis involves using a chemical, N-ethyl-N-nitrosourea (ENU), to create single base changes in the genome. ENU transfers its ethyl group to oxygen or nitrogen radicals in DNA, resulting in mis-pairing and base pair substitution. Mutant animals can be produced by injecting a male mouse with ENU, and breeding with a wild type female to produce mutant offspring. ENU mutagenesis creates a high frequency of random mutations, with approximately one base pair change in any given gene in every 200-700 gametes. Despite its high mutagenicity, the physical penetration of ENU is limited and only about 500 genes are mutated for each male and a very small number of the total mutations have an observable phenotype. Thousands of mutations typically need to be created in a single animal in order to generate one novel phenotype.

Despite recent improvements in ENU technology, mapping mutations responsible for a particular phenotype is typically difficult and time-consuming. Neutral mutations must be separated from causative mutations, via extensive breeding. ENU and cloning methods are simply inefficient for creating and mapping gene knockouts in rats for the creation of new models of human disease. Through 2007, the largest rat ENU mutagenesis project to date run by the Medical College of Wisconsin was able to produce only 9 knockout rat lines in a period of five years at an average cost of $200,000 per knockout line. Although some companies are still pursuing this strategy, the Medical College of Wisconsin has switched to a more efficient and commercially viable method using mobile DNA and CompoZr ZFN technology.

Zinc-finger and TALE Nuclease Technology

Zinc finger nucleases (ZFNs) and Transcription Activator-Like Effector Nucleases (TALENs) are engineered DNA-binding proteins that facilitate targeted editing of the genome by creating double-strand breaks in DNA at user-specified locations. Double strand breaks are important for site-specific mutagenesis in that they stimulate the cell's natural DNA-repair processes, namely homologous recombination and non-homologous end joining. When the cell uses the non-homologous end joining pathway to repair the double-strand break, the inherent inaccuracy of the repair often generates precisely targeted mutations. This results in embryos with targeted gene knockout. Standard microinjection techniques allow this technology to make knockout rats in 4–6 months. A major advantage of ZFN- and TALEN-mediated gene knockout relative to the use of mobile DNA is that a particular gene can be uniquely and specifically targeted for knockout. In contrast, knockouts made using mobile DNA technology are random and are therefore unlikely to target the gene of interest.

Mobile DNA Technology

Mobile DNA (jumping gene) technology uses retrotransposons and transposons for the production of knockout rat models. This platform technology meets all of the criteria for a successful gene knockout approach in mammals by permitting random mutagenesis directly in the germ cells (sperm and oocytes) of mammalian model organisms, including rats. Using this technology, genes are disrupted completely and in a stable manner, are knocked out at a high frequency, and are randomly disrupted throughout the entire genome. The genomic location of mutations can be easily mapped, creating a library of knockout rats for later use. Once the random knockout mutations are created, more refined mutations such as conditional mutations can be created by breeding knockout lines with rat lines expressing CRE recombinase in a tissue specific manner. Knock-ins can be produced by recombination mediated cassette exchange.

PiggyBac (PB) DNA Transposons

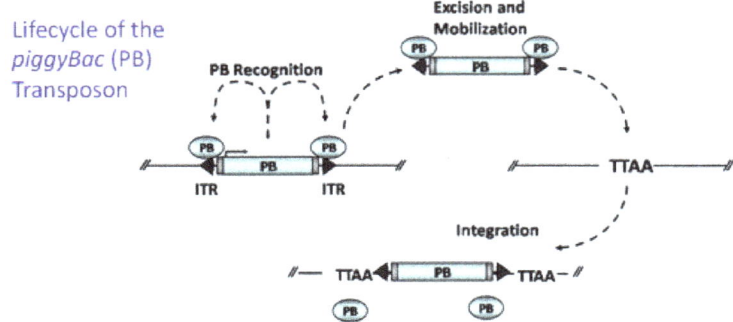

piggyBac transposon technology

piggyBac (PB) DNA transposons mobilize via a "cut-and-paste" mechanism whereby a transposase enzyme (PB transposase), encoded by the transposon itself, excises and re-integrates the transposon at other sites within the genome. PB transposase specifically recognizes PB inverted terminal repeats (ITRs) that flank the transposon; it binds to these sequences and catalyzes excision of the transposon. PB then integrates at TTAA sites throughout the genome, in a relatively random fashion. For the creation of gene trap mutations (or adapted for generating transgenic animals), the transposase is supplied in trans on one plasmid and is co-transfected with a plasmid containing donor transposon, a recombinant transposon comprising a gene trap flanked by the binding sites for the transposase (ITRs). The transposase will catalyze the excision of the transposon from the plasmid and subsequent integration into the genome. Integration within a coding region will capture the elements necessary for gene trap expression. PB possesses several ideal properties: (1) it preferentially inserts within genes (50 to 67% of insertions hit genes) (2) it exhibits no local hopping (widespread genomic coverage) (3) it is not sensitive to over-production inhibition in which elevated levels of the transposase cause decreased transposition 4) it excises cleanly from a donor site, leaving no "footprint," unlike Sleeping Beauty.

Sleeping Beauty (SB) Transposons

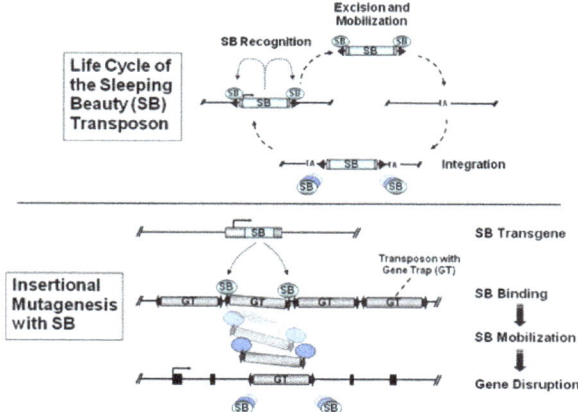

Sleeping Beauty transposon technology

The sleeping beauty (SB) transposon is a derivative of the Tc1/mariner superfamily of DNA transposons prevalent among both vertebrate and invertebrate genomes. However, endogenous DNA transposons from this family are completely inactive in vertebrate genomes. An active Tc1/mariner transposon, synthesized from alignment of inactive transposons from the salmonid subfamily of elements, was "awoken" to form the transposon named Sleeping Beauty. SB, like other DNA transposons, mobilizes itself via a cut-and-paste mechanism whereby a transposase enzyme, encoded by the transposon itself, excises and re-integrates the transposon at other sites within the genome. The 340 amino acid SB protein recognizes inverted terminal repeats (ITRs) that flank the transposon; it binds to these sequences and catalyzes excision of the transposon. SB then integrates into random sites within the genome, although some studies report very slight preferences for transcriptional units. There is also a simple requirement of a TA-dinucleotide at the target site, like all Tc1/mariner transposons.

The SB transposon is a powerful tool for insertional mutagenesis in many vertebrate species. It recently exhibited especial utility for germ line mutagenesis in both mice and rats. There are several advantages that make SB a highly attractive mutagen geared toward gene discovery: 1) it has little bias for inserting within particular genomic regions or within specific recognition sequences, 2) de novo insertions of the transposon provide a "tagged" sequence marker for rapid identification of the specific mutation by simple PCR cloning methods, 3) in vivo SB insertional mutagenesis allows multiple mutations to be quickly and easily generated in a single animal, and in a single tissue, such as an adenomatous polyp.

LINE1 (L1) Retrotransposons

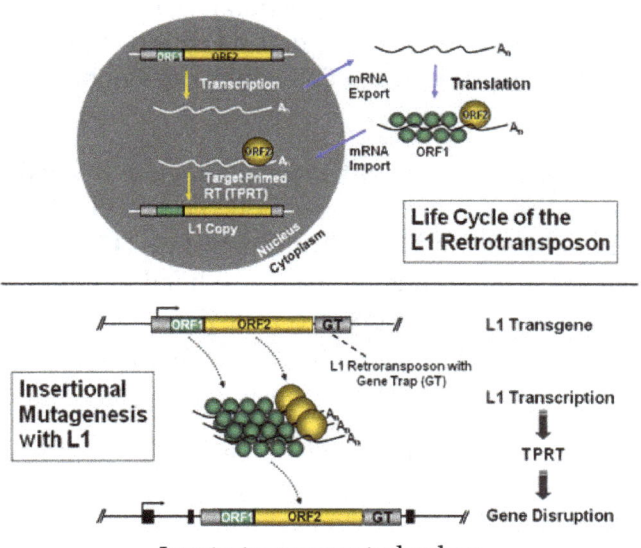

L1 retrotransposon technology

Transposons and retrotransposons are valuable tools for unbiased gene discovery as mobile pieces of DNA used for gene disruption. Retrotransposons, such as LINEs (long interspersed nuclear elements), mobilize via a "copy and paste" mechanism and are abundant in many eukaryotic species. Several L1 retrotransposons have remained active in mice and humans. L1s contain a small internal promoter within a 5' untranslated region to drive expression, two open reading frames (ORFs), and a 3' untranslated region containing sequences for polyadenylation. The two ORFs encode proteins necessary for autonomous retrotransposition; ORF1 encodes an RNA-binding protein while ORF2 encodes a protein containing endonuclease (EN) and reverse transcriptase (RT) activity, which nick a site in DNA, then produce a copy via RT. These proteins exhibit an overwhelming specificity for binding to and acting on the transcript that encodes them, enabling near exclusive mobilization of the parental L1 RNA. Using the RT activity of the ORF2 protein, the transcribed L1 RNA is copied into DNA by a process termed target primed reverse transcription (TPRT), and integrated into the genome. Integration occurs with little bias for any particular genomic region, requiring a simple consensus sequence, 5'TTTT'A-3' (along with minor variations of this sequence). Integrated L1 sequences are often truncated at the 5' end, with an average total size of 1 Kb, many containing only 3' terminal sequences.

The nature of retrotransposition endows the L1 with some unique advantages; L1 retrotransposons have an essentially unlimited supply of the insertional mutagen since it is continually transcribed from a promoter, which would be useful for applications where large numbers of mutations are needed in a single cell. L1 elements also demonstrate widespread genomic coverage, with a largely random distribution of insertions. L1 insertions at genomic sites are also irreversible, and thus any mutagenic event caused by an L1 insertion is "tagged" by L1 sequences.

Genetically Modified Fish

Genetically modified fish (GM fish) are organisms from the taxonomic clade which includes the classes Agnatha (jawless fish), Chondrichthyes (cartilaginous fish) and Osteichthyes (bony fish) whose genetic material (DNA) has been altered using genetic engineering techniques. In most cases, the aim is to introduce a new trait to the fish which does not occur naturally in the species, i.e. transgenesis.

GM fish are used in scientific research and kept as pets. They are being developed as environmental pollutant sentinels and for use in aquaculture food production. In 2015, the AquAdvantage salmon was approved by the US Food and Drug Administration (FDA) for commercial production, sale and consumption, making it the first genetically modified animal to be approved for human consumption. Some GM fish that have been created have promoters driving an over-production of "all fish" growth hormone. This results in dramatic growth enhancement in several species, including salmonids, carps and tilapias.

Critics have objected to GM fish on several grounds, including ecological concerns, animal welfare concerns and with respect to whether using them as food is safe and whether GM fish are needed to help address the world's food needs.

History and Process

The first transgenic fish were produced in China in 1985. As of 2013, approximately 50 species of fish have been subject to genetic modification. This has resulted in more than 400 fish/trait combinations. Most of the modifications have been conducted on food species, such as Atlantic salmon (*Salmo salar*), tilapia (genus) and common carp (*Cyprinus carpio*).

Generally, genetic modification entails manipulation of DNA. The process is known as cisgenesis when a gene is transferred between organisms that could be conventionally bred, or transgenesis when a gene from one species is added to a different species. Gene transfer into the genome of the desired organism, as for fish in this case, requires a vector like a lentivirus or mechanical/physical insertion of the altered genes into the nucleus of the host by means of a micro syringe or a gene gun.

Uses

Research

A zebrafish genetically modified to have long fins

Transgenic fish are used in research covering five broad areas -

- Enhancing the traits of commercially available fish

- Their use as bioreactors for the development of bio-medically important proteins

- Their use as indicators of aquatic pollutants

- Developing new non-mammalian animal models

- Functional genomics studies

Most GM fish are used in basic research in genetics and development. Two species of fish, zebrafish and medaka, are most commonly modified because they have optically clear chorions (shells), develop rapidly, the 1-cell embryo is easy to see and micro-inject with transgenic DNA, and zebrafish have the capability of regenerating their organ tissues. They are also used in drug discovery. GM zebrafish are being explored for benefits of unlocking human organ tissue diseases and failure mysteries. For instance, zebrafish are used to understand heart tissue repair and regeneration in efforts to study and discover cures for cardiovascular diseases.

Transgenic rainbow trout (*Oncorhynchus mykiss*) have been developed to study muscle development. The introduced transgene causes green fluorescence to appear in fast twitch muscle fibres early in development which persist throughout life. It has been suggested the fish might be used as indicators of aquatic pollutants or other factors which influence development.

In intensive fish farming, the fish are kept at high stocking densities. This means they suffer from frequent transmission of contagious diseases, a problem which is being addressed by GM research. Grass carp (*Ctenopharyngodon idella*) have been modified with a transgene coding for human lactoferrin, which doubles their survival rate relative to control fish after exposure to *Aeromonas* bacteria and Grass carp hemorrhage virus. Cecropin has been used in channel catfish to enhance their protection against several pathogenic bacteria by 2–4 times.

Recreation

Pets

GloFish Electric Green Tetra

GloFish is a patented technology which allows GM fish (tetra, barb, zebrafish) to express jellyfish and sea coral proteins giving the fish bright red, green or orange fluorescent colors when viewed in ultraviolet light. Although the fish were originally created and patented for scientific research at the National University of Singapore, a Texas company, Yorktown Technologies, obtained rights to market the fish as pets. They became the first genetically modified animal to become publicly available as a pet when introduced for sale in 2003. They were quickly banned for sale in California, however they are now on shelves once again in this state. As of 2013, Glofish are only sold in the US.

Other transgenic lines of pet fish include Medaka which remain transparent throughout their lives and pink body color transgenic angelfish (*Pterophyllum scalare*) and lionhead fish expressing the Acropora coral (*Acroporo millepora*) red fluorescent protein.

The ocean pout type III antifreeze protein transgene has been successfully micro-injected and expressed in goldfish. The transgenic goldfish showed higher cold tolerance compared with controls.

Food

One area of intensive research with GM fish has aimed to increase food production by modifying the expression of growth hormone (GH). The relative increases in growth differ between species. They range from a doubling in weight, to some fish that are almost 100 times heavier than the wild-type at a comparable age. This research area has resulted in dramatic growth enhancement in several species, including salmon, trout and tilapia. Other sources indicate an 11-fold and 30-fold increase in growth of salmon and mud loach, respectively, compared to wild-type fish. Transgenic fish development has reached the stage where several species are ready to be marketed in different countries, for example, GM tilapia in Cuba, GM carp in the People's Republic of China, and GM salmon in the USA and Canada. In 2014, it was reported that applications for the approval of transgenic fish as food had been made in Canada, China, Cuba and the United States.

Over-production of GH from the pituitary gland increases growth rate mainly by an increase in food consumption by the fish, but also by a 10 to 15% increase in feed conversion efficiency.

Another approach to increasing meat production in GM fish is "double muscling". This results in a phenotype similar to that of Belgian Blue cattle in rainbow trout. It is achieved by using transgenes expressing follistatin, which inhibits myostatin, and the development of two muscle layers.

AquAdvantage Salmon

In November 2015, the FDA of the USA approved the GM AquAdvantage salmon created by AquaBounty for commercial production, sale and consumption. It is the first genetically modified animal to be approved for human consumption. The fish is essentially an Atlantic salmon with a single gene complex inserted: a growth hormone regulating gene from a Chinook salmon with a promoter sequence from an ocean pout. This permits the GM salmon to produce GH year round rather than pausing for part of the year as do wild-type Atlantic salmon. The wild-type salmon takes 24 to 30 months to reach market size (4–6 kg) whereas the producers of the GM salmon claim it requires only 18 months for the GM fish to achieve this. AquaBounty argue that their GM salmon can be grown nearer to end-markets with greater efficiency (they require 25% less feed to achieve market weight) than the Atlantic salmon which are currently reared in remote coastal fish farms, thereby making it better for the environment, with recycled waste and lower transport costs.

It has been claimed that to prevent the genetically modified fish inadvertently breeding with wild salmon, all the fish will be female and reproductively sterile. However, it has also been claimed that a small percentage of the females will remain fertile. Some opponents of the GM salmon have dubbed it the "Frankenfish". Approval of the AquAdvantage fish by the FDA was opposed by a consortium of more than 20 anti-GM organisations.

Detecting Aquatic Pollution (Potential)

Several research groups have been developing GM zebrafish to detect aquatic pollution. The laboratory that developed the GloFish originally intended them to change color in the presence of pollutants, as environmental sentinels. Teams at the University of Cincinnati and Tulane University have been developing GM fish for the same purpose.

Several transgenic methods have been used to introduce target DNA into zebrafish for environmental monitoring, including micro-injection, electroporation, particle gun bombardment, liposome-mediated gene transfer, and sperm-mediated gene transfer. Micro-injection is the most commonly used method to produce transgenic zebrafish as this produces the highest survival rate.

Regulation

The regulation of genetic engineering concerns the approaches taken by governments to assess and manage the risks associated with the development and release of genetically modified crops. There are differences in the regulation of GMOs between countries, with some of the most marked differences occurring between the USA and Europe. Regulation varies in a given country depending on the intended use of the products of the genetic engineering. For example, a fish not intended for food use is generally not reviewed by authorities responsible for food safety.

The US FDA guidelines for evaluating transgenic animals define transgenic constructs as "drugs" regulated under the animal drug provisions of the Federal Food and Cosmetic Act. This classification is important for several reasons, including that it places all GM food animal permits under the jurisdiction of the FDA's Center for Veterinary Medicine (CVM) and imposes limits on what information the FDA can release to the public, and furthermore, it avoids a more open food safety review process.

The US states of Washington and Maine have imposed permanent bans on the production of transgenic fish.

Controversy

Critics have objected to use of genetic engineering per se on several grounds, including ethical concerns, ecological concerns (especially about gene flow), and economic concerns raised by the fact GM techniques and GM organisms are subject to intellectual property law. GMOs also are involved in controversies over GM food with respect to whether using GM fish as safe is safe, whether it would exacerbate or cause fish allergies, whether it should be labeled, and whether GM fish and crops are needed to address the world's food needs. These controversies have led to litigation, international trade disputes, and protests, and to restrictive regulation of commercial products in most countries.

There is much doubt among the public about genetically modified animals in general. It is believed that the acceptance of GM fish by the general public is the lowest of all GM animals used for food and pharmaceuticals.

Ethical Concerns

In transgenic fast-growing fish genetically modified for growth hormone, the mosaic founder fish vary greatly in their growth rate, reflecting the highly variable proportion and distribution of transgenic cells in their bodies. Fish with these high growth rates (and their progeny) sometimes develop a morphological abnormality similar to acromegaly in humans, exhibiting an enlarged head relative to the body and a bulging operculum. This becomes progressively worse as the fish ages. It can interfere with feeding and may ultimately cause death. According to a study commissioned by Compassion in World Farming, the abnormalities are probably a direct consequence of growth hormone over-expression and have been reported in GM coho salmon, rainbow trout, common carp, channel catfish and loach, but to a lesser extent in Nile tilapia.

In GM coho salmon (*Oncorhynchus kisutch*) there are morphological changes and changed allometry that lead to reduced swimming abilities. They also exhibit abnormal behaviour such as increased levels of activity with respect to feed-intake and swimming. Several other transgenic fish show decreased swimming ability, likely due to body shape and muscle structure.

Genetically modified triploid fish are more susceptible to temperature stress, have a higher incidence of deformities (e.g. abnormalities in the eye and lower jaw), and are less aggressive than diploids. Other welfare concerns of GM fish include increased stress under oxygen-deprived conditions caused by increased need for oxygen. It has been shown that deaths due to low levels of oxygen (hypoxia) in coho salmon are most pronounced in transgenics. It has been suggested the increased sensitivity to hypoxia is caused by the insertion of the extra set of chromosomes requiring a larger nucleus which thereby causes a larger cell overall and a reduction in the surface area to volume ratio of the cell.

Ecological Concerns

An aquaculture enterprise

Transgenic fish are usually developed in strains of near-wild origin. These have an excellent capacity for interbreeding with themselves or wild relatives and therefore possess a significant possibility for establishing themselves in nature should they escape biotic or abiotic containment measures.

A wide range of concerns about the consequences of genetically modified fish escaping have been expressed. For polyploids, these include the degree of sterility, interference with spawning, competing with resources without contributing to subsequent generations. For transgenics, the concerns include characteristics of the genotype, the function of the gene, the type of the gene, potential for causing pleiotropic effects, potential for interacting with the remainder of the genome, stability of the construct, ability of the DNA construct to transpose within or between genomes.

One study, using relevant life history data from the Japanese medaka (*Oryzias latipes*) predicts that a transgene introduced into a natural population by a small number of transgenic fish will spread as a result of enhanced mating advantage, but the reduced viability of offspring will cause eventual local extinction of both populations. GM coho salmon show greater risk-taking behaviour and better use of limited food than wild-type fish.

Transgenic coho salmon have enhanced feeding capacity and growth, which can result in a considerably larger body size (>7-fold) compared to non-transgenic salmon. When transgenic and non-transgenic salmon in the same enclosure compete for different levels of food, transgenic individuals consistently outgrow non-transgenic individuals. When food abundance is low, dominant individuals emerge, invariably transgenic, that show strong agonistic and cannibalistic behavior to cohorts and dominate the acquisition of limited food resources. When food availability is low, all groups containing transgenic salmon experience population crashes or complete extinctions, whereas groups containing only non-transgenic salmon have good (72%) survival rates. This has led to the suggestion that these GM fish will survive better than the wild-type when conditions are very poor.

Successful artificial transgenic hybridization between two species of loach (genus *Misgurnus*) has been reported, yet these species are not known to hybridize naturally.

GloFish were not considered as an environmental threat because they were less fit than normal zebrafish which are unable to establish themselves in the wild in the US.

AquAdvantage Salmon

The FDA has said the AquAdvantage Salmon can be safely contained in land-based tanks with little risk of escape into the wild, however, Joe Perry, former chair of the GM panel of the European Food Safety Authority, has been quoted as saying "There remain legitimate ecological concerns over the possible consequences if these GM salmon escape to the wild and reproduce, despite FDA assurances over containment and sterility, neither of which can be guaranteed".

AquaBounty indicates their GM salmon can not interbreed with wild fish because they are triploid which makes them sterile. The possibility of fertile triploids is one of the major short-falls of triploidy being used as a means of bio-containment for transgenic fish. However, it is estimated that 1.1% of eggs remain diploid, and therefore capable of breeding, despite the triploidy process. Others have claimed the sterility process has a failure rate of 5%. Large scale trials using normal pressure, high pressure, or high pressure plus aged eggs for transgenic coho salmon, give triploidy frequencies of only 99.8%, 97.6%, and 97.0%, respectively. AquaBounty also emphasizes that their GM salmon would not survive wild conditions due to the geographical locations where their research is conducted, as well as the locations of their farms.

The GH transgene can be transmitted via hybridization of GM AquAdvantage Salmon and the closely related wild brown trout (*Salmo trutta*). Transgenic hybrids are viable and grow more rapidly than transgenic salmon and other wild-type crosses in conditions emulating a hatchery. In stream mesocosms designed to simulate natural conditions, transgenic hybrids express competitive dominance and suppress the growth of transgenic and non-transgenic salmon by 82% and 54%, respectively. Natural levels of hybridization between these two species can be as high as 41%. Researchers examining this possibility concluded "Ultimately, we suggest that hybridization of transgenic fishes with closely related species represents potential ecological risks for wild populations and a possible route for introgression of a transgene, however low the likelihood, into a new species in nature."

An article in Slate Magazine in December 2012 by Jon Entine, Director of the Genetic Literacy Project, criticized the Obama Administration for preventing the publication of the environmental assessment (EA) of the AquAdvantage Salmon, which was completed in April 2012 and which concluded that "the salmon is safe to eat and poses no serious environmental hazards." The Slate article said that the publication of the report was stopped "after meetings with the White House, which was debating the political implications of approving the GM salmon, a move likely to infuriate a portion of its base". Within days of the article's publication and less than two months after the election, the FDA released the draft EA and opened the comment period.

GloFish

The GloFish is a patented and trademarked brand of genetically engineered fluorescent fish. A variety of different GloFish are currently on the market. Zebrafish were the first GloFish available in pet stores, and are now sold in bright red, green, orange-yellow, blue, and purple fluorescent colors. Recently "Electric Green", "Sunburst Orange", "Moonrise Pink", "Starfire Red", "Cosmic Blue", and "Galactic Purple" colored tetra (*Gymnocorymbus ternetzi*) and an "Electric Green" tiger barb (*Puntius tetrazona*) have been added to the lineup. Although not originally developed for the ornamental fish trade, it is one of the first genetically modified animals to become publicly available. It is sold only in the United States, where it remains the only genetically modified animal to be publicly available. The rights to GloFish are owned by Yorktown Technologies, the company that commercialized the fish.

History

Early Development

An ordinary Zebra Danio

The original zebrafish (or zebra danio, *Danio rerio*) is a native of rivers in India and Bangladesh. It measures three centimeters long and has gold and dark blue stripes.

In 1999, Dr. Zhiyuan Gong and his colleagues at the National University of Singapore were working with a gene that encodes the green fluorescent protein (GFP), originally extracted from a jellyfish, that naturally produced bright green fluorescence. They inserted the gene into a zebrafish embryo, allowing it to integrate into the zebrafish's genome, which caused the fish to be brightly fluorescent under both natural white light and ultraviolet light. Their goal was to develop a fish that could detect pollution by selectively fluorescing in the presence of environmental toxins. The development of the constantly fluorescing fish was the first step in this process, and the National University of Singapore filed a patent application on this work. Shortly thereafter, his team developed a line of red fluorescent zebra fish by adding a gene from a sea coral, and orange-yellow fluorescent zebra fish, by adding a variant of the jellyfish gene. Later, a team of researchers at the National Taiwan University, headed by Professor Huai-Jen Tsai, succeeded in creating a medaka (rice fish) with a fluorescent green color, which, like the zebrafish, is a model organism used in biology.

The scientists from NUS and businessmen Alan Blake and Richard Crockett from Yorktown Technologies, L.P., a company in Austin, Texas, met and a deal was signed whereby Yorktown obtained the worldwide rights to market the fluorescent zebrafish, which Yorktown subsequently branded as "GloFish". At around the same time, a separate deal was made between Taikong, the largest aquarium fish producer in Taiwan, and the Taiwanese researchers to market the green medaka in Taiwan under the name TK-1. In the spring of 2003, Taiwan became the first to authorize sales of a genetically modified organism as a pet. One hundred thousand fish were reportedly sold in less than a month at US$18.60 each. The fluorescent medaka are not GloFish, as they are not marketed by Yorktown Technologies, but instead by Taikong Corp under a different brand name.

Introduction to The United States Market

GloFish were introduced to the United States market in late 2003 by Yorktown Technologies, after two years of research. The governmental environmental risk assessment was made by the U.S. Food and Drug Administration (FDA), which has jurisdiction over all genetically modified (GM) animals, including fluorescent zebra fish, since they consider the inserted gene to be a drug.

Because tropical aquarium fish are not used for food purposes, they pose no threat to the food supply. There is no evidence that these genetically engineered zebra danio fish pose any more threat to the environment than their unmodified counterparts which have long been widely sold in the United States. In the absence of a clear risk to the public health, the FDA finds no reason to regulate these particular fish.

Marketing of the fish was met by protests from a non-governmental organization called the Center for Food Safety. They were concerned that approval of the GloFish based only on a Food and Drug Administration risk assessment would create a precedent of inadequate scrutiny of biotech animals in general. The group filed a lawsuit in US Federal District Court to block the sale of the GloFish. The lawsuit sought a court order stating that the sale of transgenic fish is subject to federal regulation beyond the FDA's charter, and as such should not be sold without more extensive approvals.

It's clear this sets a precedent for genetically engineered animals. It opens the dams to a whole host of nonfood genetically engineered organisms. That's unacceptable to us and runs counter to things the National Academy of Sciences and other scientific review boards have said, particularly when it comes to mobile GM organisms like fish and insects.

The Center for Food Safety's suit was found to be without merit and dismissed on March 30, 2005.

Developments since The Glofish Introduction

In addition to the red fluorescent zebrafish, trademarked as "Starfire Red", Yorktown Technologies released a green fluorescent zebrafish and an orange-yellow fluorescent zebrafish in mid-2006. In 2011, blue and purple fluorescent zebrafish were released. These lines of fish are trademarked as "Electric Green", "Sunburst Orange", "Cosmic Blue", and "Galactic Purple", and incorporate genes from sea coral. In 2012, Yorktown Technologies introduced a new variety of "Electric Green" GloFish, derived from a different species of fish, the black tetra. This was followed by the "Electric Green" Barb, which is a variety of tiger barb. In 2013, Yorktown Technologies introduced a "Sunburst Orange" Tetra and a "Moonrise Pink" Tetra, the first fluorescent pink fish to be marketed. This was followed in 2014 by the release of a "Starfire Red" and "Cosmic Blue" Tetra.

Despite the speculation of aquarium enthusiasts that the eggs are pressure treated to make them infertile, it has been found some GloFish are indeed fertile and will reproduce in a captive environment. However, the GloFish Fluorescent Fish License states "Intentional breeding and/or any sale, barter, or trade, of any offspring of GloFish fluorescent ornamental fish is strictly prohibited."

Sale or possession of GloFish was made illegal in California in 2002 due to a regulation that restricts genetically modified fish. The regulation was implemented before the marketing of GloFish, largely due to concern about a fast-growing biotech salmon. The regulations were lifted in 2015 due to a growing body of evidence and the findings of the Food and Drug Administration and the Florida Department of Agriculture and Consumer Services. GloFish are now legal in California for importation and commercial sale.

Canada prohibits import or sale of the fish, saying there is insufficient information to make a decision with regard to safety.

The import, sale and possession of these fish is not permitted within the European Union. On November 9, 2006, however, the Netherlands' Ministry of Housing, Spatial Planning and the Environment (VROM) found 1,400 fluorescent fish, which were sold in various aquarium shops.

In January 2009, the U.S. Food & Drug Administration formalized their recommendations for genetically engineered animals. These non-binding recommendations describe the way in which FDA regulates all GM animals, including GloFish.

Research published in 2014 assessed the environmental safety associated with GloFish. One paper concluded that there is little risk of invasiveness into the environment. A second study concluded that there is no difference in risk between GloFish and wild-type danios.

Sources of Colors

Examples of sources of fluorescent protein genes include GFP (*Aequorea victoria*, jellyfish), GFP (*Renilla reniformis*, sea pansy), dsRed (*Discosoma*, mushroom coral), eqFP611 (*Entacmaea quadricolor*, sea anemone), RTMS5 (*Montipora efflorescens*, stony coral), dronpa (*Pectiniidae*, chalice coral), KFP (*Anemonia sulcata*, Venus hair anemone), eosFP (*Lobophyllia hemprichii*, open brain coral), and dendra (*Dendronephthya*, octocoral).

In early 2014, scientists identified approximately 200 species of naturally occurring fluorescent fish, suggesting that the fluorescence trait is widespread in fish lines.

Other Experimental Uses

Fluorescent zebrafish also have been used for other experimental research. The alterations in the zebrafish's genes has given the organism the ability to fluoresce as a bio-indicator. This genetic ability has been used to detect pollution and other chemicals.

Chemicals that mimic natural estrogens have well-documented effects on the reproductive systems of vertebrates, typically acting as endocrine disruptors, and GloFish fluorescence is being used to detect levels of estrogenic chemicals. Investigators found that muscles such as the heart are more affected by estrogen than the liver. Using the GloFish may thus give insights into endocrine disrupting chemical actions.

The sentiments of aquarium retailers towards the GloFish have also been used as an indicator of the public's positive reaction to controversial agricultural biotechnologies.

Vulnerability to Predation

GloFish are more vulnerable to predation compared to the wild-type, according to a study published in 2011.

Genetically Modified Insect

The fruit-fly *Drosophila melanogaster*, often used in genetic modification studies

A genetically modified (GM) insect is an insect that has been genetically modified, either through mutagenesis, or more precise processes of transgenesis, or cisgenesis. Motivations for using GM insects include biological research purposes and genetic pest management. Genetic pest management capitalizes on recent advances in biotechnology and the growing repertoire of sequenced genomes in order to control pest populations, including insects. Insect genomes can be found in genetic databases such as NCBI, and databases more specific to insects such as FlyBase, VectorBase, and BeetleBase. There is an ongoing initiative started in 2011 to sequence the genomes of 5,000 insects and other arthropods called the i5k. Some Lepidoptera (e.g. monarch butterflies and silkworms) have been genetically modified in nature by the wasp bracovirus.

Types of Genetic Pest Management

The sterile insect technique (SIT) was developed conceptually in the 1930s and 1940s and first used in the environment in the 1950s. SIT is a control strategy where male insects are sterilized, usually by irradiation, then released to mate with wild females. If enough males are released, the females will mate with mostly sterile males and lay non-viable eggs. This causes the population of insects to crash (the abundance of insects is extremely diminished), and in some cases can lead to local eradication. Irradiation is a form of mutagenesis which causes random mutations in DNA.

Release of Insects carrying Dominant Lethals (RIDL) is a control strategy using genetically engineered insects that have (carry) a lethal gene in their genome (an organism's DNA). Lethal genes cause death in an organism, and RIDl genes only kill young insects, usually larvae or pupae. Similar to how inheritance of brown eyes is dominant to blue eyes, this lethal gene is dominant so that all offspring of the RIDL insect will also inherit the lethal gene. This lethal gene has a molecular on and off switch, allowing these RIDL insects to be reared. The lethal gene is turned off when the RIDL insects are mass reared in an insectory, and turned on when they are released into the environment. RIDL males and females are released to mate with wild males and their offspring die when they reach the larval or pupal stage because of the lethal gene. This causes the population of insects to crash. This technique is being developed for some insects and for other insects has been tested in the field. It has been used in the Grand Cayman Islands, Panama, and Brazil to control the mosquito vector of dengue, *Ae. aegypti*. It is being developed for use in diamondback moth (*Plutella xylostella),* medfly and olive fly.

Incompatible Insect Technique (IIT) - Wolbachia

Maternal Effect Dominant Embryonic Arrest (Medea)

X-Shredder

Concerns

There are concerns about using tetracycline on a routine basis for controlling the expression of lethal genes. There are plausible routes for resistance genes to develop in the bacteria within the guts of GM-insects fed on tetracycline and from there, to circulate widely in the environment. For example, antibiotic resistant genes could be spread to E.coli bacteria and into fruit by GM-Mediterranean fruit flies (*Ceratitis capitata*).

Releases

In January 2016 it was announced that in response to the Zika virus outbreak, Brazil's National Biosafety Committee approved the releases of more genetically modified Aedes aegypti mosquitos throughout their country. Previously in July 2015, Oxitec released results of a test in the Juazeiro region of Brazil, of so-called "self-limiting" mosquitoes, to fight dengue, Chikungunya and Zika viruses. They concluded that mosquito populations were reduced by about 95%.

Modified Species

Biological Research

- Fruit flies (*Drosophila melanogaster*) are model organisms used in an array of biological disciplines (i.e. neurobiology, population genetics, ecology, animal behavior, systematics, genomics, and development). Many studies done with *Drosophila* species have been foundational in their respective fields, and they remain important models for other organisms, including humans. For example, they have contributed to understanding economically important insects and researching human disease and development. Fruit flies are often preferred over other animals due to their short life cycle, reproduction rate, low maintenance requirements, and amenability to mutagenesis. They are also the model genetic organism for historical reasons, being one of the first model organism and have a high quality completed genome.

Genetic Pest Management

- Yellow fever mosquito (*Aedes aegypti*)

- Malaria mosquito (*Anopheles gambiae* and *Anopheles stephensi*)

- Pink bollworm (*Pectinophora gossypiella*)

Diamondback Moth

Diamondback moth

The diamondback moth's caterpillars gorge on cruciferous vegetables such as cabbage, broccoli, cauliflower and kale, globally costing farmers an estimated $5 billion (£3.2 million) a year worldwide. In 2015, Oxitec developed GM-diamondback moths which produce non-viable female larvae to control populations able to develop resistance to insecticides. The GM-insects were initially placed in cages for field trials. Earlier, the moth was the first crop pest to evolve resistance to DDT and eventually became resistant to 45 other insecticides. In Malaysia, the moth has become immune to all synthetic sprays. The gene is a combination of DNA from a virus and a bacterium. In an earlier study, captive males carrying the gene eradicated communities of non-GM moths. Brood sizes were similar, but female offspring died before reproducing. The gene itself disappears after a few generations, requiring ongoing introductions of GM cultivated males. Modified moths can be identified by their red glow under ultraviolet light, caused by a coral transgene.

Opponents claim that the protein made by the synthetic gene could harm non-target organisms that eat the moths. The creators claim to have tested the gene's protein on mosquitoes, fish, beetles, spiders and parasitoids without observing problems. Farmers near the test site claim that moths could endanger nearby farms' organic certification. Legal experts say that national organic standards penalize only deliberate GMO use. The creators claim that the moth does not migrate if sufficient food is available, nor can it survive winter weather.

Mediterranean Fruit Fly

The Mediterranean fruit fly is a global agricultural pest. They infest a wide range of crops (over

300) including wild fruit, vegetables and nuts, and in the process, cause substantial damage. The company Oxitec has developed GM-males which have a lethal gene that interrupts female development and kills them in a process called "pre-pupal female lethality". After several generations, the fly population diminishes as the males can no longer find mates. To breed the flies in the laboratory, the lethal gene can be "silenced" using the antibiotic tetracycline.

Mediterranean fruit fly

Opponents argue that the long-term effects of releasing millions of GM-flies are impossible to predict. Dead fly larvae could be left inside crops. Helen Wallace from Genewatch, an organisation that monitors the use of genetic technology, stated "Fruit grown using Oxitec's GM flies will be contaminated with GM maggots which are genetically programmed to die inside the fruit they are supposed to be protecting". She added that the mechanism of lethality was likely to fail in the longer term as the GM flies evolve resistance or breed in sites contaminated with tetracycline which is widely used in agriculture.

Legislation

In July 2015, the House of Lords (U.K.) Science and Technology Committee launched an inquiry into the possible uses of GM-insects and their associated technologies. The scope of the inquiry is to include questions such as "Would farmers benefit if insects were modified in order to reduce crop pests? What are the safety and ethical concerns over the release of genetically modified insects? How should this emerging technology be regulated?"

Detection of Genetically Modified Organisms

The detection of genetically modified organisms in food or feed is possible by biochemical means. It can either be qualitative, showing which genetically modified organism (GMO) is present, or quantitative, measuring in which amount a certain GMO is present. Being able to detect a GMO is an important part of GMO labeling, as without detection methods the traceability of GMOs would rely solely on documentation.

Polymerase Chain Reaction (PCR)

The polymerase chain reaction (PCR) is a biochemistry and molecular biology technique for isolating and exponentially amplifying a fragment of DNA, via enzymatic replication, without using a living organism. It enables the detection of specific strands of DNA by making millions of copies of a target genetic sequence. The target sequence is essentially photocopied at an exponential rate, and simple visualisation techniques can make the millions of copies easy to see.

The method works by pairing the targeted genetic sequence with custom designed complementary bits of DNA called primers. In the presence of the target sequence, the primers match with it and trigger a chain reaction. DNA replication enzymes use the primers as docking points and start doubling the target sequences. The process is repeated over and over again by sequential heating and cooling until doubling and redoubling has multiplied the target sequence several million-fold. The millions of identical fragments are then purified in a slab of gel, dyed, and can be seen with UV light. It is not prone to contamination. Irrespective of the variety of methods used for DNA analysis, only PCR in its different formats has been widely applied in GMO detection/analysis and generally accepted for regulatory compliance purposes. Detection methods based on DNA rely on the complementarity of two strands of DNA double helix that hybridize in a sequence-specific manner. The DNA of GMO consists of several elements that govern its functioning. The elements are promoter sequence, structural gene and stop sequence for the gene.

Quantitative Detection

Quantitative PCR (Q-PCR) is used to measure the quantity of a PCR product (preferably real-time, QRT-PCR). It is the method of choice to quantitatively measure amounts of transgene DNA in a food or feed sample. Q-PCR is commonly used to determine whether a DNA sequence is present in a sample and the number of its copies in the sample. The method with currently the highest level of accuracy is quantitative real-time PCR. QRT-PCR methods use fluorescent dyes, such as Sybr Green, or fluorophore-containing DNA probes, such as TaqMan, to measure the amount of amplified product in real time. If the targeted genetic sequence is unique to a certain GMO, a positive PCR test proves that the GMO is present in the sample.

Qualitative Detection

Whether or not a GMO is present in a sample can be tested by Q-PCR, but also by multiplex PCR. Multiplex PCR uses multiple, unique primer sets within a single PCR reaction to produce amplicons of varying sizes specific to different DNA sequences, i.e. different transgenes. By targeting multiple genes at once, additional information may be gained from a single test run that otherwise would require several times the reagents and more time to perform. Annealing temperatures for each of the primer sets must be optimized to work correctly within a single reaction, and amplicon sizes, i.e., their base pair length, should be different enough to form distinct bands when visualized by gel electrophoresis.

Event-specific vs. Construct-specific Detection

When producers, importers or authorities test a sample for the unintended presence of GMOs, they usually do not know which GMO to expect. While EU authorities prefer an event-specific approach to this problem, US authorities rely on construct-specific test schemes.

Event-specific Detection

An event-specific detection searches for the presence of a DNA sequence unique to a certain GMO, usually the junction between the transgene and the organism's original DNA. This approach is ideal to precisely identify a GMO, yet highly similar GMOs will pass completely unnoticed. Event-specific detection is PCR-based.

Construct-specific Detection

The construct-specific detection methods can either be DNA or protein based. DNA based detection looks for a part of the foreign DNA inserted in a GMO. For technical reasons, certain DNA sequences are shared by several GMOs. Protein-based methods detect the product of the transgene, for example the Bt toxin. Since different GMOs may produce the same protein, construct-specific detection can test a sample for several GMOs in one step, but is unable to tell precisely which of the similar GMOs are present. Especially in the USA, protein-based detection is used for the construct-specific approach.

Shortcomings of Current Detection Methods

Currently, it is highly unlikely that the presence of unexpected or even unknown GMOs will be detected, since either the DNA sequence of the transgene or its product, the protein, must be known for detection. In addition, even testing for known GMOs is time-consuming and costly, as current reliable detection methods can test for only one GMO at a time. Therefore, research programmes such as Co-Extra are developing improved and alternative testing methods, for example DNA microarrays.

Alternative Detection Methods

Improving PCR Based Detection

Improving PCR based detection of GMOs is a further goal of the European research programme Co-Extra. Research is now underway to develop multiplex PCR methods that can simultaneously detect many different transgenic lines. Another major challenge is the increasing prevalence of transgenic crops with stacked traits. This refers to transgenic cultivars derived from crosses between transgenic parent lines, combining the transgenic traits of both parents. One GM maize variety now awaiting a decision by the European Commission, MON863 x MON810 x NK603, has three stacked traits. It is resistant to an herbicide and to two different kinds of insect pests. Some combined testing methods could give results that would triple the actual GM content of a sample containing this GMO.

Detecting Unknown GMOs

Almost all transgenic plants contain a few common building blocks that make unknown GMOs easier to find. Even though detecting a novel gene in a GMO can be like finding a needle in a haystack, the fact that the needles are usually similar makes it much easier. To trigger gene expression, scientists couple the gene they want to add with what is known as a transcription promoter. The high-performing 35S promoter is a common feature to many GMOs. In addition, the stop signal for gene transcription in most GMOs is often the same: the NOS terminator. Researchers now compile a set of genetic sequences characteristic of GMOs. After genetic elements characteristic of GMOs are selected, methods and tools are developed for detecting them in test samples. Approaches being considered include microarrays and anchor PCR profiling.

Near Infrared Fluorescence (NIR)

Near infrared fluorescence (NIR) detection is a method that can reveal what kinds of chemicals are present in a sample based on their physical properties. By hitting a sample with near infrared light, chemical bonds in the sample vibrate and re-release the light energy at a wavelength characteristic for a specific molecule or chemical bond. It is not yet known if the differences between GMOs and conventional plants are large enough to detect with NIR imaging. Although the technique would require advanced machinery and data processing tools, a non-chemical approach could have some advantages such as lower costs and enhanced speed and mobility.

Switzerland

The Cantons of Switzerland perform tests to assess the presence of genetically modified organisms in foodstuffs. In 2008, 3% of the tested samples contained detectable amounts of GMOs. In 2012, 12% of the samples analysed contained detectable amounts of GMOs (including 2.4% of GMOs forbidden in Switzerland). Except one, all the samples tested contained less than 0.9% of GMOs; which is the threshold that impose labelling indacating the presence of GMOs.

References

- Büttner-Mainik, A.; et al. (2011). "Production of biologically active recombinant human factor H in Physcomitrella". Plant Biotechnology Journal. 9: 373–383. PMID 20723134. doi:10.1111/j.1467-7652.2010.00552.x. CS1 maint: Explicit use of et al

- M. K. Sateesh (25 August 2008). Bioethics And Biosafety. I. K. International Pvt Ltd. pp. 456–. ISBN 978-81-906757-0-3. Retrieved 27 March 2013

- "'Any idiot can do it.' Genome editor CRISPR could put mutant mice in everyone's reach". Science | AAAS. 2016-11-02. Retrieved 2016-12-02

- Klein, TM; et al. (1987). "High-velocity microprojectiles for delivering nucleic acids into living cells". Nature. 327 (6117): 70–73. Bibcode:1987Natur.327...70K. doi:10.1038/327070a0

- Classical Medicine Journal (14 April 2010). "Genetically modified cows producing human milk." Archived from the original on 6 November 2014

- Panesar, Pamit et al. (2010) Enzymes in Food Processing: Fundamentals and Potential Applications, Chapter 10, I K International Publishing House, ISBN 978-93-80026-33-6

- James, Clive (1996). "Global Review of the Field Testing and Commercialization of Transgenic Plants: 1986 to 1995" (PDF). The International Service for the Acquisition of Agri-biotech Applications. Retrieved 17 July 2010

- Lee LY, Gelvin SB (February 2008). "T-DNA binary vectors and systems". Plant Physiol. 146 (2): 325–332. OCLC 1642351. PMC 2245830. PMID 18250230. doi:10.1104/pp.107.113001

- Rahman, MA; et al. (2001). "Growth and nutritional trials on transgenic Nile tilapia containing an exogenous fish growth hormone gene". Journal of Fish Biology. 59 (1): 62–78. doi:10.1111/j.1095-8649.2001.tb02338.x. CS1 maint: Explicit use of et al.

- Biopolymer, Volume 8 Polyamides and Complex Proteinaceous Materials II, edited by S.R. Fahnestock & A. Steinbuchel, 2003 Wiley-VCH Verlag, pages 97-117 ISBN 978-3-527-30223-9

- Molteni, Megan (2017-04-12). "Florida's Orange Trees Are Dying, But a Weaponized Virus Could Save Them". WIRED. Retrieved 2017-04-17

- Walsh, Gary (April 2005). "Therapeutic insulins and their large-scale manufacture". Appl. Microbiol. Biotechnol. 67 (2): 151–159. PMID 15580495. doi:10.1007/s00253-004-1809-x

- Landrigan, Philip J.; Benbrook, Charles (2015). "GMOs, Herbicides, and Public Health". New England Journal of Medicine. New England Journal of Medicine. 373 (8): 693–5. PMID 26287848. doi:10.1056/NEJMp1505660

- Dunham, R.A.; Winn, R.N. (2014). "Chapter 11 - Production of transgenic fish". In Pinkert, C.A. Transgenic Animal Technology: A Laboratory Handbook. Elsevier. ISBN 9780323137836

- Staff (3 April 2012) Biology of HIV Archived 11 April 2014 at the Wayback Machine. National Institute of Allergy and Infectious Diseases, Retrieved 31 August 2012

- Woodard, S. L.; Woodard, J. A.; Howard, M. E. (2004). "Plant molecular farming: Systems and products". Plant Cell Reports. 22 (10): 711–720. PMID 14997337. doi:10.1007/s00299-004-0767-1

- Sanford, JC; et al. (1987). "Delivery of substances into cells and tissues using a particle bombardment process". Journal of Particulate Science and Technology. 5: 27–37. doi:10.1080/02726358708904533

- Nielsen, K. M. (2003). "Transgenic organisms—time for conceptual diversification?". Nature Biotechnology. 21 (3): 227–228. PMID 12610561. doi:10.1038/nbt0303-227

- Logan J, Edwards K, Saunders N, eds. (2009). Real-Time PCR: Current Technology and Applications. Caister Academic Press. ISBN 978-1-904455-39-4

- Goldenberg, Suzanne (25 November 2013). "Canada approves production of GM salmon eggs on commercial scale". The Guardian. Retrieved 26 November 2013

- Foster K, Foster H, Dickson JG (December 2006). "Gene therapy progress and prospects: Duchenne muscular dystrophy". Gene Ther. 13 (24): 1677–85. PMID 17066097. doi:10.1038/sj.gt.3302877

4

Genetically Modified Plants and Crops

The crops or plants that are modified by using genetic engineering techniques are known as genetically modified plants. The types of modifications used are transgenic, cisgenic and subgenic. Some of the genetically engineered plants are potato, rice, soybean, tomato, wheat and sugar beet. The topics discussed in the chapter are of great importance to broaden the existing knowledge on genetically modified plants.

Transplastomic Plant

A transplastomic plant is a genetically modified plant in which the new genes have not been inserted in the nuclear DNA but in the DNA of the chloroplasts. The major advantage of this technology is that in many plant species plastid DNA is not transmitted through pollen, which prevents gene flow from the genetically modified plant to other plants.

Transformation Technology

The most common method to transform plastids is particle bombardment: Small gold or tungsten particles are coated with DNA and shot into young plant cells or plant embryos. Some genetic material will stay in the cells and transform them. The transformation efficiency is lower than in agrobacterial mediated transformation, which is also common in plant genetic engineering, but particle bombardment is especially suitable for plastid transformation.

In order to persist and be stably maintained in the cell, a plasmid DNA molecule must contain an origin of replication, which allows it to be replicated in the cell independently of the chromosome. Because transformation usually produces a mixture of rare transformed cells and abundant non-transformed cells, a method is needed to identify the cells that have acquired the plasmid. Plasmids used in transformation experiments will usually also contain a gene giving resistance to an antibiotic (or, more recently developed resistance against a herbicide) that the intended recipient strain of bacteria is sensitive to. Selection for cells able to grow on media containing this antibiotic can then select the cells that have acquired the plasmid by transformation, as cells lacking the plasmid will be unable to grow.

Biological Containment and Agricultural Coexistence

Genetically modified plants must be safe for the environment and suitable for coexistence with conventional and organic crops. Towards such safety, a major hurdle is posed by the potential outcrossing of the transgene via pollen movement. Plastid transformation, which yields transplastomic plants in which the pollen does not contain the transgene, not only increases biosafety, but also facilitates the coexistence of genetically modified, conventional and organic agriculture. Therefore, developing such crops is a major goal of research projects such as Co-Extra and Transcontainer.

Transplastomic Tobacco

However, plastid transformation is suitable only for certain crop species, and the reliability of this method has only been proven for tobacco. Led by Ralph Bock from the Max Planck Institute of Molecular Plant Physiology in Germany, researchers studied genetically modified tobacco in which the transgene was integrated in chloroplasts. Since past literature reported contradicting figures on the reliability of this process, the Co-Extra researchers analysed more than two million seedlings and found that less than 20 in 1,000,000 inherited the transgene. In the pollen of adult plants, the rate was even lower, remaining below 3 in 1,000,000. This reduction is because some parts of the seedlings are lost during their development into mature plants.

Because tobacco has a strong tendency towards self-fertilisation, the reliability of transplastomic plants is assumed to be even higher under field conditions. Therefore, the researchers believe that only one in 100,000,000 GM tobacco plants actually would transmit the transgene via pollen. Such values are more than satisfactory to ensure coexistence. However, for GM crops used in the production of pharmaceuticals, or in other cases in which absolutely no outcrossing is permitted, the researchers recommend the combination of chloroplast transformation with other biological containment methods, such as cytoplasmic male sterility or transgene mitigation strategies.

Genetically Modified Crops

Genetically modified crops (GMCs, GM crops, or biotech crops) are plants used in agriculture, the DNA of which has been modified using genetic engineering methods. In most cases, the aim is to introduce a new trait to the plant which does not occur naturally in the species. Examples in food crops include resistance to certain pests, diseases, or environmental conditions, reduction of spoilage, or resistance to chemical treatments (e.g. resistance to a herbicide), or improving the nutrient profile of the crop. Examples in non-food crops include production of pharmaceutical agents, biofuels, and other industrially useful goods, as well as for bioremediation.

Farmers have widely adopted GM technology. Between 1996 and 2015, the total surface area of land cultivated with GM crops increased by a factor of 100, from 17,000 km² (4.2 million acres) to 1,797,000 km² (444 million acres). 10% of the world's arable land was planted with GM crops in 2010. In the US, by 2014, 94% of the planted area of soybeans, 96% of cotton and 93% of corn were genetically modified varieties. Use of GM crops expanded rapidly in developing countries, with about 18 million farmers growing 54% of worldwide GM crops by 2013. A 2014 meta-analysis concluded that GM technology adoption had reduced chemical pesticide use by 37%, increased crop yields by 22%, and increased farmer profits by 68%. This reduction in pesticide use has been ecologically beneficial, but benefits may be reduced by overuse. Yield gains and pesticide reductions are larger for insect-resistant crops than for herbicide-tolerant crops. Yield and profit gains are higher in developing countries than in developed countries.

There is a scientific consensus that currently available food derived from GM crops poses no greater risk to human health than conventional food, but that each GM food needs to be tested on a case-by-case basis before introduction. Nonetheless, members of the public are much less likely than scientists to perceive GM foods as safe. The legal and regulatory status of GM foods varies by country, with some nations banning or restricting them, and others permitting them with widely differing degrees of regulation.

However, opponents have objected to GM crops on several grounds, including environmental concerns, whether food produced from GM crops is safe, whether GM crops are needed to address the world's food needs, whether the foods are readily accessible to poor farmers in developing countries, and concerns raised by the fact these organisms are subject to intellectual property law.

Gene Transfer in Nature and Traditional Agriculture

DNA transfers naturally between organisms. Several natural mechanisms allow gene flow across species. These occur in nature on a large scale – for example, it is one mechanism for the development of antibiotic resistance in bacteria. This is facilitated by transposons, retrotransposons, proviruses and other mobile genetic elements that naturally translocate DNA to new loci in a genome. Movement occurs over an evolutionary time scale.

The introduction of foreign germplasm into crops has been achieved by traditional crop breeders by overcoming species barriers. A hybrid cereal grain was created in 1875, by crossing wheat and rye. Since then important traits including dwarfing genes and rust resistance have been introduced. Plant tissue culture and deliberate mutations have enabled humans to alter the makeup of plant genomes.

History

The first genetically modified crop plant was produced in 1982, an antibiotic-resistant tobacco plant. The first field trials occurred in France and the USA in 1986, when tobacco plants were engineered for herbicide resistance. In 1987, Plant Genetic Systems (Ghent, Belgium), founded by Marc Van Montagu and Jeff Schell, was the first company to genetically engineer insect-resistant (tobacco) plants by incorporating genes that produced insecticidal proteins from *Bacillus thuringiensis* (Bt).

The People's Republic of China was the first country to allow commercialized transgenic plants, introducing a virus-resistant tobacco in 1992, which was withdrawn in 1997.[3] The first genetically modified crop approved for sale in the U.S., in 1994, was the *FlavrSavr* tomato. It had a longer shelf life, because it took longer to soften after ripening. In 1994, the European Union approved tobacco engineered to be resistant to the herbicide bromoxynil, making it the first commercially genetically engineered crop marketed in Europe.

In 1995, Bt Potato was approved by the US Environmental Protection Agency, making it the country's first pesticide producing crop. In 1995 canola with modified oil composition (Calgene), Bt maize (Ciba-Geigy), bromoxynil-resistant cotton (Calgene), Bt cotton (Monsanto), glyphosate-resistant soybeans (Monsanto), virus-resistant squash (Asgrow), and additional delayed ripening tomatoes (DNAP, Zeneca/Peto, and Monsanto) were approved. As of mid-1996, a total of 35 approvals had been granted to commercially grow 8 transgenic crops and one flower crop (carnation), with 8 different traits in 6 countries plus the EU. In 2000, Vitamin A-enriched golden rice was developed, though as of 2016 it was not yet in commercial production. In 2013 the leaders of the three research teams that first applied genetic engineering to crops, Robert Fraley, Marc Van

Montagu and Mary-Dell Chilton were awarded the World Food Prize for improving the "quality, quantity or availability" of food in the world.

Methods

Genetically engineered crops have genes added or removed using genetic engineering techniques, originally including gene guns, electroporation, microinjection and agrobacterium. More recently, CRISPR and TALEN offered much more precise and convenient editing techniques.

Gene guns (also known as biolistics) "shoot" (direct high energy particles or radiations against) target genes into plant cells. It is the most common method. DNA is bound to tiny particles of gold or tungsten which are subsequently shot into plant tissue or single plant cells under high pressure. The accelerated particles penetrate both the cell wall and membranes. The DNA separates from the metal and is integrated into plant DNA inside the nucleus. This method has been applied successfully for many cultivated crops, especially monocots like wheat or maize, for which transformation using *Agrobacterium tumefaciens* has been less successful. The major disadvantage of this procedure is that serious damage can be done to the cellular tissue.

Agrobacterium tumefaciens-mediated transformation is another common technique. Agrobacteria are natural plant parasites, and their natural ability to transfer genes provides another engineering method. To create a suitable environment for themselves, these Agrobacteria insert their genes into plant hosts, resulting in a proliferation of modified plant cells near the soil level (crown gall). The genetic information for tumor growth is encoded on a mobile, circular DNA fragment (plasmid). When *Agrobacterium* infects a plant, it transfers this T-DNA to a random site in the plant genome. When used in genetic engineering the bacterial T-DNA is removed from the bacterial plasmid and replaced with the desired foreign gene. The bacterium is a vector, enabling transportation of foreign genes into plants. This method works especially well for dicotyledonous plants like potatoes, tomatoes, and tobacco. Agrobacteria infection is less successful in crops like wheat and maize.

Electroporation is used when the plant tissue does not contain cell walls. In this technique, "DNA enters the plant cells through miniature pores which are temporarily caused by electric pulses."

Microinjection directly injects the gene into the DNA.

Plant scientists, backed by results of modern comprehensive profiling of crop composition, point out that crops modified using GM techniques are less likely to have unintended changes than are conventionally bred crops.

In research tobacco and *Arabidopsis thaliana* are the most frequently modified plants, due to well-developed transformation methods, easy propagation and well studied genomes. They serve as model organisms for other plant species.

Introducing new genes into plants requires a promoter specific to the area where the gene is to be expressed. For instance, to express a gene only in rice grains and not in leaves, an endosperm-specific promoter is used. The codons of the gene must be optimized for the organism due to codon usage bias.

Types of Modifications

Transgenic

Transgenic plants have genes inserted into them that are derived from another species. The inserted genes can come from species within the same kingdom (plant to plant) or between kingdoms (for example, bacteria to plant). In many cases the inserted DNA has to be modified slightly in order to correctly and efficiently express in the host organism. Transgenic plants are used to express proteins like the cry toxins from *B. thuringiensis*, herbicide resistant genes, antibodies and antigens for vaccinations A study led by the European Food Safety Authority (EFSA) found also viral genes in transgenic plants.

Transgenic carrots have been used to produce the drug Taliglucerase alfa which is used to treat Gaucher's disease. In the laboratory, transgenic plants have been modified to increase photosynthesis (currently about 2% at most plants versus the theoretic potential of 9–10%). This is possible by changing the rubisco enzyme (i.e. changing C3 plants into C4 plants), by placing the rubisco in a carboxysome, by adding CO_2 pumps in the cell wall, by changing the leaf form/size. Plants have been engineered to exhibit bioluminescence that may become a sustainable alternative to electric lighting.

Cisgenic

Cisgenic plants are made using genes found within the same species or a closely related one, where conventional plant breeding can occur. Some breeders and scientists argue that cisgenic modification is useful for plants that are difficult to crossbreed by conventional means (such as potatoes), and that plants in the cisgenic category should not require the same regulatory scrutiny as transgenics.

Subgenic

Genetically modified plants can also be developed using gene knockdown or gene knockout to alter the genetic makeup of a plant without incorporating genes from other plants. In 2014, Chinese researcher Gao Caixia filed patents on the creation of a strain of wheat that is resistant to powdery mildew. The strain lacks genes that encode proteins that repress defenses against the mildew. The researchers deleted all three copies of the genes from wheat's hexaploid genome. Gao used the TALENs and CRISPR gene editing tools without adding or changing any other genes. No field trials were immediately planned. The CRISPR technique has also been used to modify white button mushrooms (*Agaricus bisporus*).

Economics

GM food's economic value to farmers is one of its major benefits, including in developing nations. A 2010 study found that Bt corn provided economic benefits of $6.9 billion over the previous 14 years in five Midwestern states. The majority ($4.3 billion) accrued to farmers producing non-Bt corn. This was attributed to European corn borer populations reduced by exposure to Bt corn, leaving fewer to attack conventional corn nearby. Agriculture economists calculated that "world surplus [increased by] $240.3 million for 1996. Of this total, the largest share (59%) went to U.S. farmers. Seed company Monsanto received the next largest share (21%), followed by US consumers (9%), the rest of the world (6%), and the germplasm supplier, Delta & Pine Land Company of Mississippi (5%)."

According to the International Service for the Acquisition of Agri-biotech Applications (ISAAA), in 2014 approximately 18 million farmers grew biotech crops in 28 countries; about 94% of the farmers were resource-poor in developing countries. 53% of the global biotech crop area of 181.5 million hectares was grown in 20 developing countries. PG Economics comprehensive 2012 study concluded that GM crops increased farm incomes worldwide by $14 billion in 2010, with over half this total going to farmers in developing countries.

Critics challenged the claimed benefits to farmers over the prevalence of biased observers and by the absence of randomized controlled trials. The main Bt crop grown by small farmers in developing countries is cotton. A 2006 review of Bt cotton findings by agricultural economists concluded, "the overall balance sheet, though promising, is mixed. Economic returns are highly variable over years, farm type, and geographical location".

In 2013 the European Academies Science Advisory Council (EASAC) asked the EU to allow the development of agricultural GM technologies to enable more sustainable agriculture, by employing fewer land, water, and nutrient resources. EASAC also criticizes the EU's "time-consuming and expensive regulatory framework" and said that the EU had fallen behind in the adoption of GM technologies.

Participants in agriculture business markets include seed companies, agrochemical companies, distributors, farmers, grain elevators and universities that develop new crops/traits and whose agricultural extensions advise farmers on best practices. According to a 2012 review based on data from the late 1990s and early 2000s, much of the GM crop grown each year is used for livestock feed and increased demand for meat leads to increased demand for GM feed crops. Feed grain usage as a percentage of total crop production is 70% for corn and more than 90% of oil seed meals such as soybeans. About 65 million metric tons of GM corn grains and about 70 million metric tons of soybean meals derived from GM soybean become feed.

In 2014 the global value of biotech seed was US$15.7 billion; US$11.3 billion (72%) was in industrial countries and US$4.4 billion (28%) was in the developing countries. In 2009, Monsanto had $7.3 billion in sales of seeds and from licensing its technology; DuPont, through its Pioneer subsidiary, was the next biggest company in that market. As of 2009, the overall Roundup line of products including the GM seeds represented about 50% of Monsanto's business.

Some patents on GM traits have expired, allowing the legal development of generic strains that include these traits. For example, generic glyphosate-tolerant GM soybean is now available. Another impact is that traits developed by one vendor can be added to another vendor's proprietary strains, potentially increasing product choice and competition. The patent on the first type of *Roundup Ready* crop that Monsanto produced (soybeans) expired in 2014 and the first harvest of off-patent soybeans occurs in the spring of 2015. Monsanto has broadly licensed the patent to other seed companies that include the glyphosate resistance trait in their seed products. About 150 companies have licensed the technology, including Syngenta and DuPont Pioneer.

Yield

In 2014, the largest review yet concluded that GM crops' effects on farming were positive. The meta-analysis considered all published English-language examinations of the agronomic and economic impacts between 1995 and March 2014 for three major GM crops: soybean, maize, and cotton. The study found that herbicide-tolerant crops have lower production costs, while for insect-resistant crops the reduced pesticide use was offset by higher seed prices, leaving overall production costs about the same.

Yields increased 9% for herbicide tolerance and 25% for insect resistant varieties. Farmers who adopted GM crops made 69% higher profits than those who did not. The review found that GM crops help farmers in developing countries, increasing yields by 14 percentage points.

The researchers considered some studies that were not peer-reviewed and a few that did not report sample sizes. They attempted to correct for publication bias, by considering sources beyond academic journals. The large data set allowed the study to control for potentially confounding variables such as fertilizer use. Separately, they concluded that the funding source did not influence study results.

Traits

GM crops grown today, or under development, have been modified with various traits. These traits include improved shelf life, disease resistance, stress resistance, herbicide resistance, pest resistance, production of useful goods such as biofuel or drugs, and ability to absorb toxins and for use in bioremediation of pollution.

Recently, research and development has been targeted to enhancement of crops that are locally important in developing countries, such as insect-resistant cowpea for Africa and insect-resistant brinjal (eggplant).

Lifetime

The first genetically modified crop approved for sale in the U.S. was the *FlavrSavr* tomato, which had a longer shelf life. It is no longer on the market.

In November 2014, the USDA approved a GM potato that prevents bruising.

In February 2015 Arctic Apples were approved by the USDA, becoming the first genetically modified apple approved for US sale. Gene silencing was used to reduce the expression of polyphenol oxidase (PPO), thus preventing enzymatic browning of the fruit after it has been sliced open. The trait was added to Granny Smith and Golden Delicious varieties. The trait includes a bacterial antibiotic resistance gene that provides resistance to the antibiotic kanamycin. The genetic engineering involved cultivation in the presence of kanamycin, which allowed only resistant cultivars to survive. Humans consuming apples do not acquire kanamycin resistance, per arcticapple.com. The FDA approved the apples in March 2015.

Nutrition

Edible Oils

Some GM soybeans offer improved oil profiles for processing or healthier eating. Camelina sativa has been modified to produce plants that accumulate high levels of oils similar to fish oils.

Vitamin Enrichment

Golden rice, developed by the International Rice Research Institute (IRRI), provides greater amounts of Vitamin A targeted at reducing Vitamin A deficiency. As of January 2016, golden rice has not yet been grown commercially in any country.

Researchers vitamin-enriched corn derived from South African white corn variety M37W, producing a 169-fold increase in Vitamin A, 6-fold increase in Vitamin C and doubled concentrations of folate. Modified Cavendish bananas express 10-fold the amount of Vitamin A as unmodified varieties.

Toxin Reduction

A genetically modified cassava under development offers lower cyanogen glucosides and enhanced protein and other nutrients (called BioCassava).

In November 2014, the USDA approved a potato, developed by J.R. Simplot Company, that prevents bruising and produces less acrylamide when fried. The modifications prevent natural, harmful proteins from being made via RNA interference. They do not employ genes from non-potato species. The trait was added to the Russet Burbank, Ranger Russet and Atlantic varieties.

Stress Resistance

Plants engineered to tolerate non-biological stressors such as drought, frost, high soil salinity, and nitrogen starvation were in development. In 2011, Monsanto's DroughtGard maize became the first drought-resistant GM crop to receive US marketing approval.

Herbicides

Glyphosate

As of 1999 the most prevalent GM trait was glyphosate-resistance. Glyphosate, (the active ingredient in Roundup and other herbicide products) kills plants by interfering with the shikimate pathway in plants, which is essential for the synthesis of the aromatic amino acids phenylalanine, tyrosine and tryptophan. The shikimate pathway is not present in animals, which instead obtain aromatic amino acids from their diet. More specifically, glyphosate inhibits the enzyme 5-enolpyruvylshikimate-3-phosphate synthase (EPSPS).

This trait was developed because the herbicides used on grain and grass crops at the time were highly toxic and not effective against narrow-leaved weeds. Thus, developing crops that could withstand spraying with glyphosate would both reduce environmental and health risks, and give an agricultural edge to the farmer.

Some micro-organisms have a version of EPSPS that is resistant to glyphosate inhibition. One of these was isolated from an *Agrobacterium* strain CP4 (CP4 EPSPS) that was resistant to glyphosate. The CP4 EPSPS gene was engineered for plant expression by fusing the 5' end of the gene to a chloroplast transit peptide derived from the petunia EPSPS. This transit peptide was used because it had shown previously an ability to deliver bacterial EPSPS to the chloroplasts of other plants. This CP4 EPSPS gene was cloned and transfected into soybeans.

The plasmid used to move the gene into soybeans was PV-GMGTO4. It contained three bacterial genes, two CP4 EPSPS genes, and a gene encoding beta-glucuronidase (GUS) from *Escherichia coli* as a marker. The DNA was injected into the soybeans using the particle acceleration method. Soybean cultivar A5403 was used for the transformation.

Bromoxynil

Tobacco plants have been engineered to be resistant to the herbicide bromoxynil.

Glufosinate

Crops have been commercialized that are resistant to the herbicide glufosinate, as well. Crops engineered for resistance to multiple herbicides to allow farmers to use a mixed group of two, three, or four different chemicals are under development to combat growing herbicide resistance.

2,4-D

In October 2014 the US EPA registered Dow's Enlist Duo maize, which is genetically modified to be resistant to both glyphosate and 2,4-D, in six states. Inserting a bacterial aryloxyalkanoate dioxygenase gene, *aad1* makes the corn resistant to 2,4-D. The USDA had approved maize and soybeans with the mutation in September 2014.

Dicamba

Monsanto has requested approval for a stacked strain that is tolerant of both glyphosate and dicamba.

Pest Resistance

Insects

Tobacco, corn, rice and many other crops have been engineered to express genes encoding for insecticidal proteins from Bacillus thuringiensis (Bt). Papaya, potatoes, and squash have been engineered to resist viral pathogens such as cucumber mosaic virus which, despite its name, infects a wide variety of plants. The introduction of Bt crops during the period between 1996 and 2005 has been estimated to have reduced the total volume of insecticide active ingredient use in the United States by over 100 thousand tons. This represents a 19.4% reduction in insecticide use.

In the late 1990s, a genetically modified potato that was resistant to the Colorado potato beetle was withdrawn because major buyers rejected it, fearing consumer opposition.

Viruses

Virus resistant papaya were developed in response to a papaya ringspot virus (PRV) outbreak in Hawaii in the late 1990s. They incorporate PRV DNA. By 2010, 80% of Hawaiian papaya plants were genetically modified.

Potatoes were engineered for resistance to potato leaf roll virus and Potato virus Y in 1998. Poor sales led to their market withdrawal after three years.

Yellow squash that were resistant to at first two, then three viruses were developed, beginning in the 1990s. The viruses are watermelon, cucumber and zucchini/courgette yellow mosaic. Squash was the second GM crop to be approved by US regulators. The trait was later added to zucchini.

Many strains of corn have been developed in recent years to combat the spread of Maize dwarf mosaic virus, a costly virus that causes stunted growth which is carried in Johnson grass and spread by aphid insect vectors. These strands are commercially available although the resistance is not standard among GM corn variants.

By-products

Drugs

In 2012, the FDA approved the first plant-produced pharmaceutical, a treatment for Gaucher's Disease. Tobacco plants have been modified to produce therapeutic antibodies.

Biofuel

Algae is under development for use in biofuels. Researchers in Singapore were working on GM jatropha for biofuel production. Syngenta has USDA approval to market a maize trademarked Enogen that has been genetically modified to convert its starch to sugar for ethanol. In 2013, the Flemish Institute for Biotechnology was investigating poplar trees genetically engineered to contain less lignin to ease conversion into ethanol. Lignin is the critical limiting factor when using wood to make bio-ethanol because lignin limits the accessibility of cellulose microfibrils to depolymerization by enzymes.

Materials

Companies and labs are working on plants that can be used to make bioplastics. Potatoes that produce industrially useful starches have been developed as well. Oilseed can be modified to produce fatty acids for detergents, substitute fuels and petrochemicals.

Bioremediation

Scientists at the University of York developed a weed (*Arabidopsis thaliana*) that contains genes from bacteria that could clean TNT and RDX-explosive soil contaminants in 2011. 16 million hectares in the USA (1.5% of the total surface) are estimated to be contaminated with TNT and RDX. However *A. thaliana* was not tough enough for use on military test grounds. Modifications in 2016 included switchgrass and bentgrass.

Genetically modified plants have been used for bioremediation of contaminated soils. Mercury, selenium and organic pollutants such as polychlorinated biphenyls (PCBs).

Marine environments are especially vulnerable since pollution such as oil spills are not containable. In addition to anthropogenic pollution, millions of tons of petroleum annually enter the marine environment from natural seepages. Despite its toxicity, a considerable fraction of petroleum oil entering marine systems is eliminated by the hydrocarbon-degrading activities of microbial communities. Particularly successful is a recently discovered group of specialists, the so-called hydrocarbonoclastic bacteria (HCCB) that may offer useful genes.

Asexual Reproduction

Crops such as maize reproduce sexually each year. This randomizes which genes get propagated to the next generation, meaning that desirable traits can be lost. To maintain a high-quality crop, some farmers purchase seeds every year. Typically, the seed company maintains two inbred varieties and crosses them into a hybrid strain that is then sold. Related plants like sorghum and gamma grass are able to perform apomixis, a form of asexual reproduction that keeps the plant's DNA intact. This trait is apparently controlled by a single dominant gene, but traditional breeding has been unsuccessful in creating asexually-reproducing maize. Genetic engineering offers another route to this goal. Successful modification would allow farmers to replant harvested seeds that retain desirable traits, rather than relying on purchased seed.

Other Modified Traits

GMO	Use	Trait	Countries approved in	First approved	Notes
Canola	Cooking oil, Margarine, Emulsifiers in packaged foods	High laurate canola	Canada	1996	
			USA	1994	
		Phytase production	USA	1998	
Carnation	Ornamental	Delayed senescence	Australia	1995	
			Norway	1998	
		Modified flower colour	Australia	1995	
			Columbia	2000	In 2014 4 ha were grown in greenhouses for export
			European Union	1998	Two events expired 2008, another approved 2007
			Japan	2004	
			Malaysia	2012	For ornamental purposes
			Norway	1997	
Maize	Animal feed, high-fructose corn syrup, corn starch	Increased lysine	Canada	2006	
			USA	2006	
		Drought tolerance	Canada	2010	
			USA	2011	
Papaya	Food	Virus resistance	China	2006	
			USA	1996	Mostly grown in Hawaii

GMO	Use	Trait	Countries approved in	First approved	Notes
Petunia	Ornamental	Modified flower colour	China	1997	
Potato	Food	Virus resistance	Canada	1999	
			USA	1997	
	Industrial	Modified starch	USA	2014	
Rose	Ornamental	Modified flower colour	Australia	2009	Surrendered renewal
			Colombia	2010	Greenhouse cultivation for export only.
			Japan	2008	
			USA	2011	
Soybean	Animal feed / Soybean oil	Increased oleic acid production	Argentina	2015	
			Canada	2000	
			USA	1997	
		Stearidonic acid production	Canada	2011	
			USA	2011	
Squash	Food	Virus resistance	USA	1994	
Sugar Cane	Food	Drought tolerance	Indonesia	2013	Environmental certificate only
Tobacco	Cigarettes	Nicotine reduction	USA	2002	

Development

The number of USDA-approved field releases for testing grew from 4 in 1985 to 1,194 in 2002 and averaged around 800 per year thereafter. The number of sites per release and the number of gene constructs (ways that the gene of interest is packaged together with other elements)—have rapidly increased since 2005. Releases with agronomic properties (such as drought resistance) jumped from 1,043 in 2005 to 5,190 in 2013. As of September 2013, about 7,800 releases had been approved for corn, more than 2,200 for soybeans, more than 1,100 for cotton, and about 900 for potatoes. Releases were approved for herbicide tolerance (6,772 releases), insect resistance (4,809), product quality such as flavor or nutrition (4,896), agronomic properties like drought resistance (5,190), and virus/fungal resistance (2,616). The institutions with the most authorized field releases include Monsanto with 6,782, Pioneer/DuPont with 1,405, Syngenta with 565, and USDA's Agricultural Research Service with 370. As of September 2013 USDA had received proposals for releasing GM rice, squash, plum, rose, tobacco, flax, and chicory.

Farming Practices

Bt Resistance

Constant exposure to a toxin creates evolutionary pressure for pests resistant to that toxin. Over-reliance on glyphosate and a reduction in the diversity of weed management practices allowed the spread of glyphosate resistance in 14 weed species/biotypes in the US.

One method of reducing resistance is the creation of refuges to allow nonresistant organisms to survive and maintain a susceptible population.

To reduce resistance to Bt crops, the 1996 commercialization of transgenic cotton and maize came with a management strategy to prevent insects from becoming resistant. Insect resistance management plans are mandatory for Bt crops. The aim is to encourage a large population of pests so that any (recessive) resistance genes are diluted within the population. Resistance lowers evolutionary fitness in the absence of the stressor (Bt). In refuges, non-resistant strains outcompete resistant ones.

With sufficiently high levels of transgene expression, nearly all of the heterozygotes (S/s), i.e., the largest segment of the pest population carrying a resistance allele, will be killed before maturation, thus preventing transmission of the resistance gene to their progeny. Refuges (i. e., fields of non-transgenic plants) adjacent to transgenic fields increases the likelihood that homozygous resistant (s/s) individuals and any surviving heterozygotes will mate with susceptible (S/S) individuals from the refuge, instead of with other individuals carrying the resistance allele. As a result, the resistance gene frequency in the population remains lower.

Complicating factors can affect the success of the high-dose/refuge strategy. For example, if the temperature is not ideal, thermal stress can lower Bt toxin production and leave the plant more susceptible. More importantly, reduced late-season expression has been documented, possibly resulting from DNA methylation of the promoter. The success of the high-dose/refuge strategy has successfully maintained the value of Bt crops. This success has depended on factors independent of management strategy, including low initial resistance allele frequencies, fitness costs associated with resistance, and the abundance of non-Bt host plants outside the refuges.

Companies that produce Bt seed are introducing strains with multiple Bt proteins. Monsanto did this with Bt cotton in India, where the product was rapidly adopted. Monsanto has also; in an attempt to simplify the process of implementing refuges in fields to comply with Insect Resistance Management(IRM) policies and prevent irresponsible planting practices; begun marketing seed bags with a set proportion of refuge (non-transgenic) seeds mixed in with the Bt seeds being sold. Coined "Refuge-In-a-Bag" (RIB), this practice is intended to increase farmer compliance with refuge requirements and reduce additional labor needed at planting from having separate Bt and refuge seed bags on hand. This strategy is likely to reduce the likelihood of Bt-resistance occurring for corn rootworm, but may increase the risk of resistance for lepidopteran corn pests, such as European corn borer. Increased concerns for resistance with seed mixtures include partially resistant larvae on a Bt plant being able to move to a susceptible plant to survive or cross pollination of refuge pollen on to Bt plants that can lower the amount of Bt expressed in kernels for ear feeding insects.

Herbicide Resistance

Best management practices (BMPs) to control weeds may help delay resistance. BMPs include applying multiple herbicides with different modes of action, rotating crops, planting weed-free seed, scouting fields routinely, cleaning equipment to reduce the transmission of weeds to other fields, and maintaining field borders. The most widely planted GMOs are designed to tolerate herbicides. By 2006 some weed populations had evolved to tolerate some of the same herbicides. Palmer amaranth is a weed that competes with cotton. A native of the southwestern US, it traveled east and was first found resistant to glyphosate in 2006, less than 10 years after GM cotton was introduced.

Plant Protection

Farmers generally use less insecticide when they plant Bt-resistant crops. Insecticide use on corn farms declined from 0.21 pound per planted acre in 1995 to 0.02 pound in 2010. This is consistent with the decline in European corn borer populations as a direct result of Bt corn and cotton. The establishment of minimum refuge requirements helped delay the evolution of Bt resistance. However, resistance appears to be developing to some Bt traits in some areas.

Tillage

By leaving at least 30% of crop residue on the soil surface from harvest through planting, conservation tillage reduces soil erosion from wind and water, increases water retention, and reduces soil degradation as well as water and chemical runoff. In addition, conservation tillage reduces the carbon footprint of agriculture. A 2014 review covering 12 states from 1996 to 2006, found that a 1% increase in herbicde-tolerant (HT) soybean adoption leads to a 0.21% increase in conservation tillage and a 0.3% decrease in quality-adjusted herbicide use.

Regulation

The regulation of genetic engineering concerns the approaches taken by governments to assess and manage the risks associated with the development and release of genetically modified crops. There are differences in the regulation of GM crops between countries, with some of the most marked differences occurring between the USA and Europe. Regulation varies in a given country depending on the intended use of each product. For example, a crop not intended for food use is generally not reviewed by authorities responsible for food safety.

Production

In 2013, GM crops were planted in 27 countries; 19 were developing countries and 8 were developed countries. 2013 was the second year in which developing countries grew a majority (54%) of the total GM harvest. 18 million farmers grew GM crops; around 90% were small-holding farmers in developing countries.

Country	2013– GM planted area (million hectares)	Biotech crops
USA	70.1	Maize, Soybean, Cotton, Canola, Sugarbeet, Alfalfa, Papaya, Squash
Brazil	40.3	Soybean, Maize, Cotton
Argentina	24.4	Soybean, Maize, Cotton
India	11.0	Cotton
Canada	10.8	Canola, Maize, Soybean, Sugarbeet
Total	175.2	----

The United States Department of Agriculture (USDA) reports every year on the total area of GMO varieties planted in the United States. According to National Agricultural Statistics Service, the states published in these tables represent 81–86 percent of all corn planted area, 88–90 percent of all soybean planted area, and 81–93 percent of all upland cotton planted area (depending on the year).

Global estimates are produced by the International Service for the Acquisition of Agri-biotech Applications (ISAAA) and can be found in their annual reports, "Global Status of Commercialized Transgenic Crops".

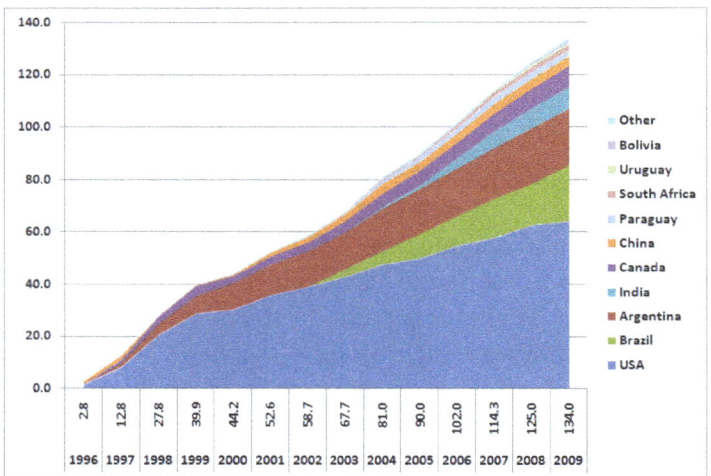

Land area used for genetically modified crops by country (1996–2009), in millions of hectares. In 2011, the land area used was 160 million hectares, or 1.6 million square kilometers.

Farmers have widely adopted GM technology. Between 1996 and 2013, the total surface area of land cultivated with GM crops increased by a factor of 100, from 17,000 square kilometers (4,200,000 acres) to 1,750,000 km² (432 million acres). 10% of the world's arable land was planted with GM crops in 2010. As of 2011, 11 different transgenic crops were grown commercially on 395 million acres (160 million hectares) in 29 countries such as the USA, Brazil, Argentina, India, Canada, China, Paraguay, Pakistan, South Africa, Uruguay, Bolivia, Australia, Philippines, Myanmar, Burkina Faso, Mexico and Spain. One of the key reasons for this widespread adoption is the perceived economic benefit the technology brings to farmers. For example, the system of planting glyphosate-resistant seed and then applying glyphosate once plants emerged provided farmers with the opportunity to dramatically increase the yield from a given plot of land, since this allowed them to plant rows closer together. Without it, farmers had to plant rows far enough apart to control post-emergent weeds with mechanical tillage. Likewise, using Bt seeds means that farmers do not have to purchase insecticides, and then invest time, fuel, and equipment in applying them. However critics have disputed whether yields are higher and whether chemical use is less, with GM crops.

In the US, by 2014, 94% of the planted area of soybeans, 96% of cotton and 93% of corn were genetically modified varieties. Genetically modified soybeans carried herbicide-tolerant traits only, but maize and cotton carried both herbicide tolerance and insect protection traits (the latter largely Bt protein). These constitute "input-traits" that are aimed to financially benefit the producers, but may have indirect environmental benefits and cost benefits to consumers. The Grocery Manufacturers of America estimated in 2003 that 70–75% of all processed foods in the U.S. contained a GM ingredient.

Europe grows relatively few genetically engineered crops with the exception of Spain, where one fifth of maize is genetically engineered, and smaller amounts in five other countries. The EU had a 'de facto' ban on the approval of new GM crops, from 1999 until 2004. GM crops are now regulated by the EU. In 2015, genetically engineered crops are banned in 38 countries worldwide, 19 of them in Europe. Developing countries grew 54 percent of genetically engineered crops in 2013.

In recent years GM crops expanded rapidly in developing countries. In 2013 approximately 18 million farmers grew 54% of worldwide GM crops in developing countries. 2013's largest increase was in Brazil (403,000 km² versus 368,000 km² in 2012). GM cotton began growing in India in 2002, reaching 110,000 km² in 2013.

According to the 2013 ISAAA brief: "...a total of 36 countries (35 + EU-28) have granted regulatory approvals for biotech crops for food and/or feed use and for environmental release or planting since 1994... a total of 2,833 regulatory approvals involving 27 GM crops and 336 GM events (NB: an "event" is a specific genetic modification in a specific species) have been issued by authorities, of which 1,321 are for food use (direct use or processing), 918 for feed use (direct use or processing) and 599 for environmental release or planting. Japan has the largest number (198), followed by the U.S.A. (165, not including "stacked" events), Canada (146), Mexico (131), South Korea (103), Australia (93), New Zealand (83), European Union (71 including approvals that have expired or under renewal process), Philippines (68), Taiwan (65), Colombia (59), China (55) and South Africa (52). Maize has the largest number (130 events in 27 countries), followed by cotton (49 events in 22 countries), potato (31 events in 10 countries), canola (30 events in 12 countries) and soybean (27 events in 26 countries).

Controversy

GM foods are controversial and the subject of protests, vandalism, referenda, legislation, court action and scientific disputes. The controversies involve consumers, biotechnology companies, governmental regulators, non-governmental organizations and scientists. The key areas are whether GM food should be labeled, the role of government regulators, the effect of GM crops on health and the environment, the effects of pesticide use and resistance, the impact on farmers, and their roles in feeding the world and energy production.

There is a scientific consensus that currently available food derived from GM crops poses no greater risk to human health than conventional food, but that each GM food needs to be tested on a case-by-case basis before introduction. Nonetheless, members of the public are much less likely than scientists to perceive GM foods as safe. The legal and regulatory status of GM foods varies by country, with some nations banning or restricting them, and others permitting them with widely differing degrees of regulation.

No reports of ill effects have been documented in the human population from GM food. Although GMO labeling is required in many countries, the United States Food and Drug Administration does not require labeling, nor does it recognize a distinction between approved GMO and non-GMO foods.

Advocacy groups such as Center for Food Safety, Union of Concerned Scientists, Greenpeace and the World Wildlife Fund claim that risks related to GM food have not been adequately examined and managed, that GMOs are not sufficiently tested and should be labelled, and that regulatory authorities and scientific bodies are too closely tied to industry. Some studies have claimed that genetically modified crops can cause harm; a 2016 review that reanalyzed the data from six of these studies found that their statistical methodologies were flawed and did not demonstrate harm, and said that conclusions about GMO crop safety should be drawn from "the totality of the evidence... instead of far-fetched evidence from single studies".

Genetically Modified Canola

Genetically modified canola is a genetically modified crop. The first strain, Roundup Ready canola, was developed by Monsanto for tolerance to glyphosate, the active ingredient in the commonly used herbicide Roundup.

Genetic Modification

Glyphosate is a broad-spectrum herbicide, which is used to kill weeds and grasses which are known to compete with commercial crops grown around the world. The first product came onto the market in the 1970s under the name 'Roundup'. Plants which are exposed to glyphosate are unable to produce aromatic amino acids and in turn die.

To produce the Roundup Ready canola, two genes were introduced into the canola genome. One is a gene derived from the common soil bacterium *Agrobacterium* strain CP4, that encodes for the *EPSPS* enzyme. The other is a gene from the *Ochrobactrum anthropi* strain LBAA, which encodes for the enzyme glyphosate oxidase (*GOX*). The CP4 *EPSPS* enzyme imparts high tolerance to glyphosate, so the plants can still create aromatic amino acids even after glyphosate is applied. *GOX* helps break down glyphosate within the plant.

Regulation

Genetically modified crops undergo a significant amount of regulation throughout the world.

For a GM crop to be approved for release in the US, it must be assessed by the Animal and Plant Health Inspection Service (APHIS) agency within the US Department of Agriculture (USDA) and may also be assessed by the Food and Drug Administration (FDA) and the Environmental protection agency (EPA), depending on the intended use. The USDA evaluates the plant's potential to become a weed. The FDA regulates crops used as food or animal feed. In Canada, the largest producer of GM canola, GM crops are regulated by Health Canada, under the Food and Drugs Act, and the Canadian Food Inspection Agency are responsible for evaluating the safety and nutritional value of genetically modified foods. Environmental assessments of biotechnology-derived plants are carried out by the CFIA's Plant Biosafety Office (PBO). In Australia Roundup Ready Canola was approved for commercial production in 2003 by the Gene Technology Regulator after undergoing approximately 400 tests and studies to determine it was safe. Food Standards Australia New Zealand also approved this product as being safe for human consumption in the same year.

Controversy

Controversy exists over the use of food and other goods derived from genetically modified crops instead of from conventional crops, and other uses of genetic engineering in food production. The dispute involves consumers, biotechnology companies, governmental regulators, nongovernmental organizations, and scientists. The key areas of controversy related to GMO foods are whether they should be labeled, the role of government regulators, the objectivity of scientific research and publication, the effect of GM crops on health and the environment, the effect on pesticide resistance, the impact of GM crops for farmers, and the role of GM crops in feeding the world population.

There is a scientific consensus that currently available food derived from GM crops poses no greater risk to human health than conventional food, but that each GM food needs to be tested on a case-by-case basis before introduction. Nonetheless, members of the public are much less likely than scientists to perceive GM foods as safe. The legal and regulatory status of GM foods varies by country, with some nations banning or restricting them, and others permitting them with widely differing degrees of regulation.

Advocacy groups such as Greenpeace, the Non-GMO Project, and Organic Consumers Association say that risks of GM food have not been adequately identified and managed, and have questioned the objectivity of regulatory authorities. They have expressed concerns about the objectivity of regulators and rigor of the regulatory process, about contamination of the non-GM food supply, about effects of GMOs on the environment and nature, and about the consolidation of control of the food supply in companies that make and sell GMOs.

Resistances Problems

Due to the heavy reliance of glyphosate in agriculture, resistance to this chemical is a problem and is prevalent throughout Australia, the USA, and Canada.

Roundup canola has also emerged as a weed in other crops due to its glyphosate resistance. This is due to canola seed being able to be dormant in the soil for up to 10 years. In California, it has become a significant problem in this way because of the restrictions on phenoxy herbicides being used in the state due to crops such as the sensitivity of cotton and grapes to this chemical.

Genetically Engineered Potato

Amflora potatoes, modified to produce pure amylopectin starch

A genetically engineered potato is a potato that has had its genes modified, using genetic engineering. Goals of modification include introducing pest resistance, tweaking the amounts of certain chemicals produced by the plant, and to prevent browning or bruising of the tubers. Varieties modified to produce large amounts of starches may be approved for industrial use only, not for food.

Currently Marketed Varieties

Innate

The genetically modified Innate potato was approved by the USDA in 2014 and the FDA in 2015. The cultivar was developed by J. R. Simplot Company. It is designed to resist blackspot bruising, browning and to contain less of the amino acid asparagine that turns into acrylamide during the frying of potatoes. Acrylamide is a probable human carcinogen, so reduced levels of it in fried potato foods is desirable. The 'Innate' name comes from the fact that this variety does not contain any genetic material from other species (the genes used are "innate" to potatoes) and uses RNA interference to switch off genes. Simplot hopes that not including genes from other species will assuage consumer fears about biotechnology.

The "Innate" potato is not a single cultivar, rather, it is a group of potato varieties that have had the same genetic alterations applied using the same process. Five different potato varieties have been transformed, creating "innate" versions of the varieties, with all of the original traits, plus the engineered ones. Ranger Russet, Russet Burbank,and Atlantic potatoes have all been transformed by Simplot, as well as two proprietary varieties. Modifications of each variety involved two transformations, one for each of the two new traits, thus there was a total of ten transformation events in developing the different Innate varieties.

McDonald's is a major consumer of potatoes in the US. The Food and Water Watch has petitioned the company to reject the newly marketed Innate potatoes. McDonald's has announced that they have ruled out using Innate.

Previously Marketed Varieties

NewLeaf

In 1995, Monsanto introduced the NewLeaf variety of potato which was their first genetically modified crop. It was designed to resist attack from the Colorado potato beetle due to the insertion of Bt toxin producing genes from the bacterium *Bacillus thuringiensis*. The insect-resistant potatoes found only a small market, and Monsanto discontinued the sale of seed in 2001.

Used in Industry

Amflora

'Amflora' (also known as EH92-527-1) was a cultivar developed by BASF Plant Science for production of pure amylopectin starch for processing into waxy potato starch. It was approved for industrial applications in the European Union market on 2 March 2010 by the European Commission, but was withdrawn from the EU market in January 2012 due to a lack of acceptance from farmers and consumers.

Unmarketed Varieties

A modified Désirée potato was developed in the 1990s by biochemist John Gatehouse at Cambridge Agricultural Genetics (later renamed Axis Genetics) and had gone through two years of field trials at Rothamsted Experimental Station. The potatoes were modified to express the *Galanthus nivalis agglutinin* (GNA) gene from the *Galanthus* (snowdrop) plant, which caused them to produce GNA lectin protein that is toxic to some insects. This variety of potatoes is the one which was involved in the Pusztai affair.

In 2014, a team of British scientists published a paper about three-year field trial showing that another genetically modified version of the Désirée cultivar can resist infection after exposure to late blight, one of the most serious diseases of potatoes. They developed this potato for blight resistance by inserting a gene (Rpi-vnt1.1), into the DNA of Désirée potatoes. This gene, which conferred the resistance to blight, was isolated from a wild relative of potatoes, *Solanum venturii*, which is a native of South America.

In 2017 scientists in Bangladesh developed their own variety of blight resistant GM potato.

Genetically Modified Rice

Rice plants being used for genetic modification

Genetically modified rice are rice strains that have been genetically modified (also called genetic engineering). Rice plants have been modified to increase micronutrients such as vitamin A, accelerate photosynthesis, tolerate herbicides, resist pests, increase grain size, generate nutrients, flavours or produce human proteins.

The natural movement of genes across species, often called horizontal gene transfer or lateral gene transfer, can also occur with rice through gene transfer mediated by natural vectors. Transgenic events between rice and *Setaria* millet have been identified. The cultivation and use of genetically modified varieties of rice remains controversial and is not approved in some countries.

History

In 2000, the first two GM rice varieties both with herbicide-resistance, called LLRice60 and LL-Rice62, were approved in the United States. Later, these and other types of herbicide-resistant GM rice were approved in Canada, Australia, Mexico and Colombia. However, none of these approvals triggered commercialization. Reuters reported in 2009 that China had granted biosafety approval to GM rice with pest resistance, but that strain was not commercialized. As of December 2012 GM rice was not widely available for production or consumption. Advocates claim that since rice is a staple crop across the world, improvements have potential to alleviate hunger, malnutrition and poverty.

Traits

Herbicide Resistance

In 2000-2001 Monsanto researched adding glyphosate tolerance to rice but did not attempt to bring a variety to market. Bayer's line of herbicide resistant rice is known as LibertyLink. Liberty-Link rice is resistant to glufosinate (the active chemical in Liberty herbicide). Bayer CropScience is attempting to get their latest variety (LL62) approved for use in the EU. The strain is approved for use in the US but is not in large-scale use. Clearfield rice was bred by selection from variations created in environments known to cause accelerated rates of mutations. This variety tolerates imidazole herbicides. It was bred by traditional breeding techniques that are not considered to be genetic engineering. Clearfield is also crossbred with higher yielding varieties to produce an overall hardier plant.

Nutritional Value

Golden Rice grains (right) compared to regular rice grains (left)

Golden rice with higher concentrations of Vitamin A was originally created by Ingo Potrykus and his team. This genetically modified rice is capable of producing beta-carotene in the endosperm (grain) which is a precursor for vitamin A. Syngenta was involved in the early development of Golden Rice and held some intellectual property that it donated to non-profit groups including the International Rice Research Institute (IRRI) to develop on a non-profit basis. The scientific details of the rice were first published in *Science Magazine* in 2000.

The World Health Organization stated that iron deficiency affects 30% of the world's population. Research scientists from the Australian Centre for Plant Functional Genomics (ACPFG) and IRRI to are working to increase the amount of iron in rice. They have modified three populations of rice by over expressing the genes OsNAS1, OsNAS2 or OsNAS3. The research team found that nicotianamine, iron, and zinc concentration levels increased in all three populations relative to controls.

Ventria Bioscience uses a proprietary system known as Express Tec for producing recombinant human proteins in rice grains. Their most notable variety produces human Lactoferrin and Lysozyme. These two proteins are produced naturally in human breast milk and are used globally in infant formula and rehydration products.

Golden Rice plants being grown in greenhouse

Pest Resistance

BT rice is modified to express the cryIA(b) gene of the *Bacillus thuringiensis* bacterium. The gene confers resistance to a variety of pests including the rice borer through the production of endo-toxins. The Chinese Government is doing field trials on insect resistant cultivars. The benefit of BT rice is that farmers do not need to spray their crops with pesticides to control fungal, viral, or bacterial pathogens. Conventional rice is sprayed three to four times per growing season to control pests. Other benefits include increased yield and revenue from crop cultivation. China approved the rice for large-scale use as of 2009.

Allergy Resistance

Researchers in Japan are attempting to develop hypoallergenic rice cultivars. Researchers are trying to repress the formation of allergen AS-Albumin.

Japanese researchers tested genetically modified rice on macaque monkeys that would prevent allergies to cedar pollen, which causes hay fever. Cedar allergy symptoms include itchy eyes, sneezing and other serious allergic reactions. The modified rice contains seven proteins within cedar pollen to block these symptoms. Takaiwa is conducting human clinical trials with this riceHuman blood protein.

Human serum albumin (HSA) is a blood protein in human blood plasma. It is used to treat severe burns, liver cirrhosis and hemorrhagic shock. It is also used in donated blood and is in short supply around the world. In China, scientists modified brown rice as a cost-effective way to produce HSA protein. The Chinese scientists put recombinant HSA protein promoters into 25 rice plants using *Agrobacterium*. Out of the 25 plants, nine contained the HSA protein. The genetically modified brown rice had the same amino acid sequence as HSA. They called this protein *Oryza sativa* recombinant HSA. The modified rice was transparent.

C4 Photosynthesis

In 2015 a consortium of 12 laboratories in eight countries developed a cultivar that displayed a rudimentary form of C4 photosynthesis (C4P) to boost growth by capturing carbon dioxide and concentrated it in specialized leaf cells. C4P is the reason corn and sugarcane grow so rapidly. Engineering C4 photosynthesis into rice could increase yields per hectare by roughly 50 percent. The current cultivar still relies primarily on C3 photosynthesis. To get them to completely adopt C4P, the plants must produce specialized cells in a precise arrangement: one set of cells to capture the carbon dioxide and to surround other cells that concentrate it. Some (possibly dozens of) genes involved in producing these cells remain to be identified. Other C3P crops that could exploit such knowledge include wheat, potatoes, tomatoes, apples and soybeans.

Controversy

Controversies surround genetically modified organisms on several levels, including ethics, environmental impact, food safety, product labeling, role in meeting world food requirements, intellectual property, and role in industrial agriculture.

Legal Issues

US

In the summer of 2006, the USDA detected trace amounts of LibertyLink variety 601 in rice shipments ready for export. LL601 was not approved for food purposes. Bayer applied for deregulation of LL601 in late July and the USDA granted deregulation status in November 2006. The contamination led to a dramatic dip in rice futures markets with losses to farmers who grew rice for export. Approximately 30 percent of rice production and 11,000 farmers in Arkansas, Louisiana, Mississippi, Missouri and Texas were affected. In June 2011 Bayer agreed to pay 750 million dollars in damages and lost harvests. Japan and Russia suspended rice imports from the U.S., while Mexico and the European Union imposed strict testing. The contamination occurred between 1998 and 2001. The exact cause of the contamination was not discovered.

China

The Chinese government does not issue commercial usage licenses for genetically modified rice. All GM rice is approved for research only. Pu, et al., stated that rice engineered to produce human blood protein (HSA) requires a lot of modified rice to be grown. This raised environmental safety concerns about gene flow. They argued that this would not be a problem because rice is a self-pollinating crop, and their test showed less than 1% of the modified gene transferred in pollination. Another study suggested that insect-mediated gene flow may be higher than previously assumed.

Genetically Modified Soybean

A genetically modified soybean is a soybean (*Glycine max*) that has had DNA introduced into it using genetic engineering techniques. In 1994 the first genetically modified soybean was introduced to the U.S. market, by Monsanto. In 2014, 90.7 million hectares of GM soy were planted worldwide, 82% of the total soy cultivation area.

Examples of Transgenic Soybeans

The genetic makeup of a soybean gives it a wide variety of uses, thus keeping it in high demand. First, manufacturers only wanted to use transgenics to be able to grow more soy at a minimal cost to meet this demand, and to fix any problems in the growing process, but they eventually found they could modify the soybean to contain healthier components, or even focus on one aspect of the soybean to produce in larger quantities. These phases became known as the first and second generation of genetically modified (GM) foods. As Peter Celec describes, "benefits of the first generation of GM foods were oriented towards the production process and companies, the second generation of GM foods offers, on contrary, various advantages and added value for the consumer", including "improved nutritional composition or even therapeutic effects."

Roundup Ready Soybean

Roundup Ready Soybeans (The first variety was also known as GTS 40-3-2 (OECD UI: MON-04032-6)) are a series of genetically engineered varieties of glyphosate-resistant soybeans produced by Monsanto.

Glyphosate kills plants by interfering with the synthesis of the essential amino acids phenylalanine, tyrosine and tryptophan. These amino acids are called "essential" because animals cannot make them; only plants and micro-organisms can make them and animals obtain them by eating plants.

Plants and microorganisms make these amino acids with an enzyme that only plants and lower organisms have, called 5-enolpyruvylshikimate-3-phosphate synthase (EPSPS). EPSPS is not present in animals, which instead obtain aromatic amino acids from their diet.

Roundup Ready Soybeans express a version of EPSPS from the CP4 strain of the bacteria, *Agrobacterium tumefaciens*, expression of which is regulated by an enhanced 35S promoter (E35S) from cauliflower mosaic virus (CaMV), a chloroplast transit peptide (CTP4) coding sequence from Petunia hybrida, and a nopaline synthase (nos 3') transcriptional termination element from Agrobacterium tumefaciens. The plasmid with EPSPS and the other genetic elements mentioned above was inserted into soybean germplasm with a gene gun by scientists at Monsanto and Asgrow. The patent on the first generation of Roundup Ready soybeans expired in March 2015.

History

First approved commercially in the United States during 1994, GTS 40-3-2 was subsequently introduced to Canada in 1995, Japan and Argentina in 1996, Uruguay in 1997, Mexico and Brazil in 1998, and South Africa in 2001.

Detection

GTS 40-3-2 can be detected using both nucleic acid and protein analysis methods.

Generic GMO Soybeans

Following expiration of Monsanto's patent on the first variety of glyphosate-resistant Roundup Ready soybeans, development began on glyphosate-resistant "generic" soybeans. The first variety, developed at the University of Arkansas Division of Agriculture, came on the market in 2015. With a slightly lower yield than newer Monsanto varieties, it costs about half as much, and seeds can be saved for subsequent years. According to its creator it is adapted to conditions in Arkansas. Several other varieties are being bred by crossing the original variety of Roundup Ready soybeans with other soybean varieties.

Stacked Traits

Monsanto developed a glyphosate-resistant soybean that also expresses Cry1Ac protein from Bacillus thuringiensis and the glyphosate-resistance gene, which completed the Brazilian regulatory process in 2010.

Genetic Modification to Improve Soybean Oil

Soy has been genetically modified to improve the quality of soy oil. Soy oil has a fatty acid profile that makes it susceptible to oxidation, which makes it rancid, and this has limited its usefulness to the food industry. Genetic modifications increased the amount of oleic acid and stearic acid and decreased the amount of linolenic acid. By silencing, or knocking out, the delta 9 and delta 12 desaturases. DuPont Pioneer created a high oleic fatty acid soybean with levels of oleic acid greater than 80%, and started marketing it in 2010.

Regulation

The regulation of genetic engineering concerns the approaches taken by governments to assess and manage the risks associated with the development and release of genetically modified crops. There are differences in the regulation of GM crops between countries, with some of the most marked differences occurring between the USA and Europe. Soy beans are allowed a Maximum Residue Limit of glyphosate of 20 mg/Kg for international trade. Regulation varies in a given country depending on the intended use of the products of the genetic engineering. For example, a crop not intended for food use is generally not reviewed by authorities responsible for food safety.

Controversy

There is a scientific consensus that currently available food derived from GM crops poses no greater risk to human health than conventional food, but that each GM food needs to be tested on a case-by-case basis before introduction. Nonetheless, members of the public are much less likely than scientists to perceive GM foods as safe. The legal and regulatory status of GM foods varies by country, with some nations banning or restricting them, and others permitting them with widely differing degrees of regulation.

A 2010 study found that in the United States, GM crops also provide a number of ecological benefits.

Critics have objected to GM crops on several grounds, including ecological concerns, and economic concerns raised by the fact these organisms are subject to intellectual property law. GM crops also are involved in controversies over GM food with respect to whether food produced from GM crops is safe and whether GM crops are needed to address the world's food needs. These controversies have led to litigation, international trade disputes, and protests, and to restrictive legislation in most countries.

Genetically Modified Sugar Beet

A genetically modified sugar beet is a sugar beet that has been genetically engineered by the direct manipulation of its genome using biotechnology. Commercialized GM sugar beets make use of a glyphosate-resistance modification developed by Monsanto and KWS Saat. These glyphosate-resistant beets, also called 'Roundup Ready' sugar beets, were developed by 2000, but not commercialized until 2007. Sugar beets are allowed a Maximum Residue Limit of glyphosate of 15 mg/Kg for international trade. As of 2016, GM sugar beets are grown in the United States and Canada. In the United States, they play an important role in domestic sugar production. Studies have concluded the sugar from glyphosate-resistant sugar beets has the same nutritional value as sugar from conventional (non-GMO) sugar beets.

More than 1 million acres of sugar beets are cultivated annually in the United States, with a market value at harvest exceeding $1 billion. GM sugar beets are grown by more than 95 percent of the nation's sugar beet farmers. The United States imports 30% of its sugar, while the remaining 70% is extracted from domestically grown sugar beets and sugarcane. Of the domestically grown sugar crops, over half of the extracted sugar is derived from sugar beets, and the rest from sugarcane.

The widespread adoption of glyphosate-resistant sugar beet has decreased the demand for migrant workers, who have historically been employed as seasonal workers to pull weeds on conventional sugar beet farms in the United States. The glyphosate sprayed on GM beet fields significantly reduces weed growth.

According to Monsanto, more than 37,000 acres of Roundup Ready sugar beet have been planted in Canada.

History

Glyphosate-resistant sugar beets were initially developed by Monsanto and KWS Saat prior to 2000. Food companies raised concerns about consumer response to GM-sourced sugar, and as a result seed companies chose not to pursue commercialization at that time.

In 2005, the US Department of Agriculture-Animal and Plant Health Inspection Service (US-DA-APHIS) deregulated glyphosate-resistant sugar beets after it conducted an environmental assessment and determined glyphosate-resistant sugar beets were highly unlikely to become a plant pest. Sugar from glyphosate-resistant sugar beets has been approved for human and animal consumption in multiple countries, but commercial production of biotech beets has been approved only in the United States and Canada.

In 2007, GM sugar beets were commercialized and GM seed sold in the United States. In 2008/2009, 60% of the sugar beets grown in the US were GM. By 2009/2010, the percentage of GM beets had grown to 95%.

In August 2010, commercial planting of GM sugar beets was suspended following a lawsuit and US district court revocation of their approval. In February 2011, the USDA-APHIS allowed GM sugar beet planting under a set of monitoring and handling requirements. In July 2012, after completing an environmental impact assessment and a plant pest risk assessment, the USDA deregulated Monsanto's Roundup Ready sugar beets.

In 2015, The Hershey Company, historically a major buyer of beet sugar, switched to cane sugar for many products due to consumer concern about GMOs.

Controversies

Litigation Over Commercial Regulation

On January 23, 2008, the Center for Food Safety, the Sierra Club, and the Organic Seed Alliance and High Mowing Seeds filed a lawsuit against USDA-APHIS regarding their decision to deregulate glyphosate-resistant sugar beets in 2005. The organizations expressed concerns regarding glyphosate-resistant sugar beets' ability to potentially cross-pollinate with conventional sugar beets.

On September 21, 2009, U.S. District Judge Jeffrey S. White, US District Court for the Northern District of California, ruled that USDA-APHIS had violated federal law in deregulating glyphosate-resistant sugar beets without adequately evaluating the environmental and socioeconomic impacts of allowing commercial production. The USDA estimated a sugar shortage would cost consumers $2.972 billion in 2011.

On August 13, 2010, Judge White revoked the deregulation of glyphosate-resistant sugar beets and declared it unlawful for growers to plant glyphosate-resistant sugar beets in the spring of 2011. As a result of this ruling, growers were permitted to harvest and process their crop at the end of the 2010 growing season, yet a ban on new plantings was enacted. After the ruling, glyphosate-resistant sugar beets could not be planted until USDA-APHIS filed an environmental impact statement (EIS), the purpose of which is to determine if environmental issues have negative effects on humans and the environment, and it may take two to three years to complete the study. After the EIS is completed, USDA-APHIS may petition to deregulate glyphosate-resistant sugar beets.

After Judge White's ruling, USDA-APHIS prepared an environmental assessment seeking partial deregulation of glyphosate-resistant sugar beets. The assessment was filed based on a request received from Monsanto and KWS SSAT AG, a German seed company. Both companies, as well as the sugar beet industry employees and growers, believed a sugar shortage would occur if glyphosate-resistant sugar beets could not be planted. As a response to this concern, USDA-APHIS developed three options in the environmental assessment to address the concerns of environmentalists, as well as those raised by the industry. The first option was to not plant glyphosate-resistant sugar beets until the EIS was completed. The second option was to allow growers to plant glyphosate-resistant sugar beets if they obtained a USDA-APHIS permit and followed specific mandates. Under the third and final option, glyphosate-resistant sugar beets would be partially deregulated, but monitored by Monsanto and KWS SSAT AG. USDA-APHIS preferred the second option. They placed the environmental assessment in the Federal Register on November 4, 2010, and received public comment for 30 days. In November 2010, in response to a suit by the original parties, Judge White ordered the destruction of plantings of genetically modified sugar beets developed by Monsanto after ruling previously that the USDA had illegally approved the biotech crop. In February 2011, a federal appeals court for the Northern district of California in San Francisco overturned the ruling, concluding, "The Plaintiffs have failed to show a likelihood of irreparable injury. Biology, geography, field experience, and permit restrictions make irreparable injury unlikely."

On February 4, 2011, the USDA-APHIS announced glyphosate-resistant sugar beets had been partially deregulated and growers would be allowed to plant seed from spring 2011 until an EIS is completed. USDA-APHIS developed requirements that growers must follow if handling glyphosate-resistant sugar beets and will monitor growers throughout the partial deregulation period. The requirements are classified into categories which include planting glyphosate-resistant sugar beets for seed production, planting for sugar production, and transporting sugar beets across state lines. Failure to follow the requirements set by USDA-APHIS may result in civil or criminal charges and destruction of the crop. In July 2012, after completing an environmental impact assessment and a plant pest risk assessment the USDA deregulated Monsanto's Roundup Ready sugar beets.

Potential for Cross-Pollination

More recently, some growers of chard seed in Oregon have raised concerns about the possibility of cross-pollination with GM sugar beets via windblown pollen.

Glyphosate-Resistant Weeds

As with other glyphosate-resistant crops, GM sugar beet farming may contribute to the growing number of glyphosate-resistant weeds. GM Corn, GM soybeans and GM cotton are grown on many times the acreage devoted to sugar beets and these crops are most affected.

Genetically Modified Tomato

A genetically modified tomato, or transgenic tomato, is a tomato that has had its genes modified, using genetic engineering. The first commercially available genetically modified food was a tomato engineered to have a longer shelf life (the Flavr Savr). Currently there are no genetically modified tomatoes available commercially, but scientists are developing tomatoes with new traits like increased resistance to pests or environmental stresses. Other projects aim to enrich tomatoes with substances that may offer health benefits or be more nutritious. As well as aiming to produce novel crops, scientists produce genetically modified tomatoes to understand the function of genes naturally present in tomatoes.

Plant physiologist Athanasios Theologis with tomatoes that contain the bioengineered ACC synthase gene

Wild tomatoes are small, green and largely unappetizing, but after centuries of breeding there are now thousands of varieties grown worldwide. Agrobacterium-mediated genetic engineering techniques were developed in the late 1980s that could successfully transfer genetic material into the nuclear genome of tomatoes. Genetic material can also be inserted into a tomato cell's chloroplast and chromoplast plastomes using biolistics. Tomatoes were the first food crop with an edible fruit where this was possible.

Examples

Delayed Ripening

Tomatoes have been used as a model organism to study the fruit ripening of climacteric fruit. To understand the mechanisms involved in the process of ripening, scientists have genetically engineered tomatoes.

In 1994, the Flavr Savr became the first commercially grown genetically engineered food to be granted a license for human consumption. A second copy of the tomato gene *polygalacturonase* was inserted into the tomato genome in the antisense direction. The polygalacturonase enzyme degrades pectin, a component of the tomato cell wall, causing the fruit to soften. When the antisense gene is expressed it interferes with the production of the polygalacturonase enzyme, delaying the ripening process. The Flavr Savr failed to achieve commercial success and was withdrawn from the market in 1997. Similar technology, but using a truncated version of the polygalacturonase gene, was used to make a tomato paste.

DNA Plant Technology (DNAP), Agritope and Monsanto developed tomatoes that delayed ripening by preventing the production of ethylene, a hormone that triggers ripening of fruit. All three tomatoes inhibited ethylene production by reducing the amount of 1-aminocyclopropane-1-carboxylic acid (ACC), the precursor to ethylene. DNAP's tomato, called Endless Summer, inserted a truncated version of the *ACC synthase* gene into the tomato that interfered with the endogenous *ACC synthase*. Monsanto's tomato was engineered with the *ACC deaminase* gene from the soil bacterium *Pseudomonas chlororaphis* that lowered ethylene levels by breaking down ACC. Agritope introduced an S-adenosylmethionine hydrolase (SAMase) encoding gene derived from the *E. coli* bacteriophage T3, which reduced the levels of S-adenosylmethionine, a precursor to ACC. Endless Summer was briefly tested in the marketplace, but patent arguments forced its withdrawal.

Scientists in India have delayed the ripening of tomatoes by silencing two genes encoding N-glycoprotein modifying enzymes, α-mannosidase and β-D-N-acetylhexosaminidase. The fruits produced were not visibly damaged after being stored at room temperature for 45 days, whereas unmodified tomatoes had gone rotten. In India, where 30% of fruit is wasted before it reaches the market due to a lack of refrigeration and poor road infrastructure, the researchers hope genetic engineering of the tomato may decrease wastage.

Environmental Stress Tolerance

Abiotic stresses like frost, drought and increased salinity are a limiting factor to the growth of tomatoes. While no genetically modified stress-tolerant plants are currently commercialised, transgenic approaches have been researched. An early tomato was developed that contained an antifreeze gene (*afa3*) from the winter flounder with the aim of increasing the tomato's tolerance to frost, which became an icon in the early years of the debate over genetically modified foods, especially in relation to the perceived ethical dilemma of combining genes from different species. This tomato gained the moniker "fish tomato". The antifreeze protein was found to inhibit ice recrystallization in the flounder blood, but had no effect when expressed in transgenic tobacco. The resulting tomato was never commercialized, possibly because the transgenic plant did not perform well in its frost-tolerance or other agronomic characteristics.

Other genes from various species have been inserted into the tomato with the hope of increasing their resistance to various environmental factors. A gene from rice (*Osmyb4*), which codes for a transcription factor, that was shown to increase cold and drought tolerance in transgenic *Arabidopsis thaliana* plants, was inserted into the tomato. This resulted in increased drought tolerance, but did not appear to have any effect on cold tolerance. Overexpressing a vacuolar Na^+/H^+ antiport (*AtNHX1*) from *A. thaliana* lead to salt accumulating in the leaves of the plants, but not in the fruit and allowed them to grow more in salt solutions than wildtype plants. They were the first salt-tolerant, edible plants ever created. Tobacco osmotic genes overexpressed in tomatoes produced plants that held a higher water content than wildtype plants increasing tolerance to drought and salt stress.

Pest Resistance

The insecticidal toxin from the bacterium *Bacillus thuringiensis* has been inserted into a tomato plant. When field tested they showed resistance to the tobacco hornworm (*Manduca sexta*), tomato fruitworm (*Heliothis zea*), the tomato pinworm (*Keiferia lycopersicella*) and the tomato fruit borer (*Helicoverpa armigera*). A 91-day feeding trial in rats showed no adverse effects, but the Bt tomato has never been commercialised. Tomatoes resistant to a root knot nematode have been created by inserting a cysteine proteinase inhibitor gene from taro. A chemically synthesised *ceropin B* gene, usually found in the giant silk moth (*Hyalophora cecropia*), has been introduced into tomato plants and in vivo studies show significant resistance to bacterial wilt and bacterial spot. When the cell wall proteins, polygalacturonase and expansin are prevented from being produced in fruits, they are less susceptible to the fungus *Botrytis cinerea* than normal tomatoes. Pest resistant tomatoes can reduce the ecological footprint of tomato production while at the same time increase farm income.

Improved Nutrition

Tomatoes have been altered in attempts to add nutritional content. In 2000, the concentration of pro-vitamin A was increased by adding a bacterial gene encoding phytoene desaturase, although the total amount of carotenoids remained equal. The researchers admitted at the time that it had no prospect of being grown commercially due to the anti-GM climate. Sue Meyer of the pressure group Genewatch, told *The Independent* that she believed, "If you change the basic biochemistry, you could alter the levels of other nutrients very important for health". More recently, scientists created blue tomatoes that have increased the production of anthocyanin, an antioxidant in tomatoes in several ways. One group added a transcription factor for the production of anthocyanin from *Arabidopsis thaliana* whereas another used transcription factors from snapdragon (*Antirrhinum*). When the snapdragon genes where used, the fruits had similar anthocyanin concentrations to blackberries and blueberries. The inventors of the GMO blue tomato using snapdragon genes, Jonathan Jones and Cathie Martin of the John Innes Centre, founded a company called Norfolk Plant Sciences to commercialize the blue tomato. They partnered with a company in Canada called New Energy Farms to grow a large crop of blue tomatoes, from which to create juice to test in clinical trials on the way to obtaining regulatory approval.

Another group has tried to increase the levels of isoflavone, known for its potential cancer preventive properties, by introducing the soybean *isoflavone synthase* into tomatoes.

Improved Taste

When geraniol synthase from lemon basil (*Ocimum basilicum*) was expressed in tomato fruits under a fruit-specific promoter, 60% of untrained taste testers preferred the taste and smell of the transgenic tomatoes. The fruits contained around half the amount of lycopene, reducing the health benefits of eating them.

Vaccines

Tomatoes (along with potatoes, bananas and other plants) are being investigated as vehicles for delivering edible vaccines. Clinical trials have been conducted on mice using tomatoes expressing antibodies or proteins that stimulate antibody production targeted to norovirus, hepatitis B, rabies, HIV, anthrax and respiratory syncytial virus. Korean scientists are looking at using the tomato to express a vaccine against Alzheimer's disease. Hilary Koprowski, who was involved in the development of the polio vaccine, led a group of researchers in developing a tomato expressing a recombinant vaccine to SARS.

Basic Research

Tomatoes are used as a model organism in scientific research and they are frequently genetically modified to further understanding of particular processes. Tomatoes have been used as a model in map-based cloning, where transgenic plants must be created to prove that a gene has been successfully isolated. The plant peptide hormone, systemin was first identified in tomato plants and genetic modification has been used to demonstrate its function, by adding antisense genes to silence the native gene or by adding extra copies of the native gene.

Genetically Modified Wheat

Genetically modified wheat is wheat that has been genetically engineered by the direct manipulation of its genome using biotechnology. As of 2015, no GM wheat is grown commercially, although many field tests have been conducted.

Background

Wheat is a natural hybrid derived from interspecies breeding. It is theorized that wheat's ancestors (*Triticum monococcum, Aegilops speltoides*, and *Aegilops tauschii*, all diploid grasses) hybridized naturally over millennia somewhere in West Asia, to create natural polyploid hybrids, the best known of which are common wheat and durum wheat.

Wheat (*Triticum* spp.) is an important domesticated grass used worldwide for food. Its evolution has been influenced by human intervention since the dawn of agriculture.

Interspecies transfer of genes continued to occur in farmers fields during the shift from the Paleolithic diet to the diet adopted by humans following the Neolithic Revolution, or first green revolution. During the transition from a hunter-gatherer social structure to more agrarian societies, humans began to cultivate wheat and further transform it for their needs. Thus, the social and cultural roots of humans and the development of wheat have intertwined since before recorded history.

This process resulted in various wheat species that are grown for specific purposes and climates. In 1873 Wilson cross-pollinated rye and wheat to create triticale. Further transformations using cytogenic hybridization techniques enabled Norman Borlaug, father of the second Green Revolution, to develop wheat species (the semidwarf varieties) that would grow in harsh environments.

Recombinant DNA techniques were developed in the 1980s, work began on creating the first transgenic wheat, coincident with the third Green Revolution. Of the three most important cereals in the world (corn, rice and wheat), wheat was the last to be transformed by transgenic, biolistic methods in 1992, and by Agrobacterium methods in 1997. Unlike corn and rice, its widespread use in the human diet has faced cultural resistance.

Field Trials and Approvals

As of 2013, 34 field trials of GM wheat have taken place in Europe and 419 have taken place in the US. Modifications tested include those to create resistance to herbicides, create resistance to insects and to fungal pathogens (especially fusarium) and viruses, tolerance to drought and resistance to salinity and heat, increased and decreased content of glutenin, improved nutrition (higher protein content, increased heat stability of the enzyme phytase, increased content of water-soluble dietary fiber, increased lysine content), improved qualities for use as biofuel feedstock, production of drugs via pharming and yield increases.

As of 2015, no GM wheat had been approved for release anywhere in the world.

Monsanto's MON 71800

The transgenic wheat that was furthest developed was Monsanto's MON 71800, which is glyphosate-resistant via a CP4/maize EPSPS gene. Monsanto received approval from the FDA for its use in food, but withdrew its EPA application in 2004, so the product was never marketed. It also received approval for use as food in Colombia.

Studies conducted by Monsanto showed that its nutritional components are equivalent to non-transgenic commercially available wheat, and animal studies that have used MON 71800 for feed have confirmed this. Environmental Risk assessments have been conducted by Monsanto, and government regulatory agencies have approved its use in food.

However, farmers were worried about the potential loss of markets in Europe and Asia due to public refusal of the end-product, so Monsanto withdrew its EPA application for Roundup-Ready Wheat.

In 2010 Monsanto's partner in India, Mahyco, announced that it planned to seek approval to market GM wheat in India in the next three to five years.

Escape of GM Wheat Seed

In 1999 scientists in Thailand claimed they discovered glyphosate-resistant wheat in a grain shipment from the Pacific Northwest of the United States, even though transgenic wheat had never been approved for sale and was only ever grown in test plots. No one could explain how the transgenic wheat got into the food supply.

In May 2013 a strain of genetically-engineered glyphosate-resistant wheat was found on a farm in Oregon. Extensive testing confirmed the wheat as a variety – MON71800. The wheat had been developed by Monsanto but never been approved or marketed after the company had tested it between 1998 and 2005. The unexplained presence of this type of wheat presents a problem to wheat growers when buyers demand GMO-free wheat. Japan subsequently suspended import of soft white wheat from the United States. A Kansas farmer sued Monsanto over the release, saying it had caused the price of wheat grown in the US to fall. Monsanto suggested that the presence of this wheat was likely an act of sabotage. On Jun 14, 2013, the USDA announced: "As of today, USDA has neither found nor been informed of anything that would indicate that this incident amounts to more than a single isolated incident in a single field on a single farm. All information collected so far shows no indication of the presence of GE wheat in commerce." As of August 30, 2013, while the source of the GM wheat remained unknown, Japan, South Korea and Taiwan had all resumed placing orders, and the disruption of the export market was minimal.

The investigation was closed in 2014 after the APHIS had exhausted all leads but had not found any evidence that the wheat had entered commercial supply.

Regulation

The regulation of genetic engineering concerns the approaches taken by governments to assess and manage the risks associated with the development and release of genetically modified crops. There are differences in the regulation of GM crops between countries, with some of the most marked differences occurring between the USA and Europe. Regulation varies in a given country depending on the intended use of the products of the genetic engineering. For example, a crop not intended for food use is generally not reviewed by authorities responsible for food safety.

Controversy

Critics have objected to GM crops per se on several grounds, including ecological concerns, and economic concerns raised by the fact these organisms are subject to intellectual property law. GM crops also are involved in controversies over GM food with respect to whether food produced from GM crops is safe and whether GM crops are needed to address the world's food needs. See the genetically modified food controversies article for discussion of issues about GM crops and GM food. These controversies have led to litigation, international trade disputes, and protests, and to restrictive legislation in a limited number of countries.

References

- Shrawat, A.; Lörz, H. (2006). "Agrobacterium-mediated transformation of cereals: a promising approach crossing barriers". Plant biotechnology journal. 4 (6): 575–603. PMID 17309731. doi:10.1111/j.1467-7652.2006.00209.x

- Martins VAP (2008). "Genomic Insights into Oil Biodegradation in Marine Systems". Microbial Biodegradation: Genomics and Molecular Biology. Caister Academic Press. ISBN 978-1-904455-17-2

- ISAAA 2013 Annual Report Executive Summary, Global Status of Commercialized Biotech/GM Crops: 2013 ISAAA Brief 46-2013, Retrieved 6 August 2014

- Predieri, S. (2001). "Mutation induction and tissue culture in improving fruits". Plant Cell, Tissue and Organ Culture. 64 (2/3): 185–210. doi:10.1023/A:1010623203554

- Koornneef, M.; Meinke, D. (2010). "The development of Arabidopsis as a model plant". The Plant journal : for cell and molecular biology. 61 (6): 909–921. PMID 20409266. doi:10.1111/j.1365-313X.2009.04086.x

- James, Clive (2007). "Executive Summary". Global Status of Commercialized Biotech/GM Crops: 2007. ISAAA Briefs. 37. The International Service for the Acquisition of Agri-biotech Applications (ISAAA). ISBN 978-1-892456-42-7

- Final Report of the PABE research project (December 2001). "Public Perceptions of Agricultural Biotechnologies in Europe". Commission of European Communities. Retrieved February 24, 2016

- Bruening, G.; Lyons, J. M. (2000). "The case of the FLAVR SAVR tomato". California Agriculture. 54 (4): 6–7. doi:10.3733/ca.v054n04p6

- Tennille, Tracy (February 13, 2015). "First Genetically Modified Apple Approved for Sale in U.S.". Wall Street Journal. Retrieved October 3, 2016

- "A decade of EU-funded GMO research (2001–2010)" (PDF). Directorate-General for Research and Innovation. Biotechnologies, Agriculture, Food. European Commission, European Union. 2010. ISBN 978-92-79-16344-9. doi:10.2777/97784. Retrieved February 8, 2016

- Brookes, Graham and Barfoot, Peter (May 2012) GM crops: global socio-economic and environmental impacts 1996-2010 PG Economics Ltd. UK, Retrieved 3 January 2012

- Walmsley, A.; Arntzen, C. (2000). "Plants for delivery of edible vaccines". Current Opinion in Biotechnology. 11 (2): 126–9. PMID 10753769. doi:10.1016/S0958-1669(00)00070-7

- Dong, H. Z.; Li, W. J. (2007). "Variability of Endotoxin Expression in Bt Transgenic Cotton". Journal of Agronomy & Crop Science. 193: 21–9. doi:10.1111/j.1439-037X.2006.00240.x

- "A decade of EU-funded GMO research (2001–2010)" (PDF). Directorate-General for Research and Innovation. Biotechnologies, Agriculture, Food. European Commission, European Union. 2010. ISBN 978-92-79-16344-9. doi:10.2777/97784. Retrieved February 8, 2016

- G. R. Heck; et al. (1 January 2005). "Development and Characterization of a CP4 EPSPS-Based, Glyphosate-Tolerant Corn Event" (Free full text). Crop Sci. 45 (1): 329–339. doi:10.2135/cropsci2005.0329

- Kilman, Scott (21 March 2001). "Monsanto's Genetically Modified Potatoes Find Slim Market, Despite Repelling Bugs". Wall Street Journal. Retrieved 23 April 2015

- Carrington, Damien (19 January 2012) GM microbe breakthrough paves way for large-scale seaweed farming for biofuels The Guardian. Retrieved 12 March 2012

- Culpepper, Stanley A; et al. (2006). "Glyphosate-resistant Palmer amaranth (Amaranthus palmeri) confirmed in Georgia". Weed Science. 54 (4): 620–626. doi:10.1614/ws-06-001r.1

- "GMO Update: US-EU Biotech Dispute; EU Regulations; Thailand". International Centre for Trade and Sustainable Development. Retrieved 5 January 2010

- Krimsky, Sheldon (2015). "An Illusory Consensus behind GMO Health Assessment" (PDF). Science, Technology, & Human Values. 40 (6): 1–32. doi:10.1177/0162243915598381

- Douglas Munier; Kent Brittan; UC Farm Advisors (December 2010). "Roundup ready canola as a resistant weed". Western Farm Press. Retrieved 8 November 2013

5

An Integrated Study of Applied Genetics

Genetic testing recognizes variations that occur in chromosomes, genes or proteins. These tests are done to study genes to understand the possibility of specific diseases or disorders that can occur in the organism in population ecology and stock quality in agriculture. In order to completely understand genetics it is necessary to understand the processes related to it. The following chapter elucidates the varied processes and mechanisms associated with this area of study.

Genetic Testing

Genetic testing, also known as DNA testing, allows the determination of bloodlines and the genetic diagnosis of vulnerabilities to inherited diseases. In agriculture, a form of genetic testing known as progeny testing can be used to evaluate the quality of breeding stock. In population ecology, genetic testing can be used to track genetic strengths and vulnerabilities of species populations.

In humans, genetic testing can be used to determine a child's parentage (genetic mother and father) or in general a person's ancestry or biological relationship between people. In addition to studying chromosomes to the level of individual genes, genetic testing in a broader sense includes biochemical tests for the possible presence of genetic diseases, or mutant forms of genes associated with increased risk of developing genetic disorders.

Genetic testing identifies changes in chromosomes, genes, or proteins. The variety of genetic tests has expanded throughout the years. In the past, the main genetic tests searched for abnormal chromosome numbers and mutations that lead to rare, inherited disorders. Today, tests involve analyzing multiple genes to determine the risk of developing specific diseases or disorders, with the more common diseases consisting of heart disease and cancer. The results of a genetic test can confirm or rule out a suspected genetic condition or help determine a person's chance of developing or passing on a genetic disorder. Several hundred genetic tests are currently in use, and more are being developed.

Because genetic mutations can directly affect the structure of the proteins they code for, testing for specific genetic diseases can also be accomplished by looking at those proteins or their metabolites, or looking at stained or fluorescent chromosomes under a microscope.

Types

Genetic testing is "the analysis of chromosomes (DNA), proteins, and certain metabolites in order to detect heritable disease-related genotypes, mutations, phenotypes, or karyotypes for clinical purposes." It can provide information about a person's genes and chromosomes throughout life. Available types of testing include:

- Cell-free fetal DNA (cffDNA) testing is a non-invasive (for the fetus) test. It is performed on a sample of venous blood from the mother, and can provide information about the fetus early in pregnancy. As of 2015 it is the most sensitive and specific screening test for Down syndrome.

- Newborn screening: Newborn screening is used just after birth to identify genetic disorders that can be treated early in life. A blood sample is collected with a heel prick from the newborn 24–48 hours after birth and sent to the lab for analysis. In the United States, newborn screening procedure varies state by state, but all states by law test for at least 21 disorders. If abnormal results are obtained, it does not necessarily mean the child has the disorder. Diagnostic tests must follow the initial screening to confirm the disease. The routine testing of infants for certain disorders is the most widespread use of genetic testing—millions of babies are tested each year in the United States. All states currently test infants for phenylketonuria (a genetic disorder that causes mental illness if left untreated) and congenital hypothyroidism (a disorder of the thyroid gland). People with PKU do not have an enzyme needed to process the amino acid phenylalanine, which is responsible for normal growth in kids and normal protein use throughout their lifetime. If there is a build-up of too much phenylalanine, brain tissue can be damaged, causing developmental delay. Newborn screening can detect the presence of PKU, allowing kids to get put on a special diet right away to avoid the effects of the disorder.

- Diagnostic testing: Diagnostic testing is used to diagnose or rule out a specific genetic or chromosomal condition. In many cases, genetic testing is used to confirm a diagnosis when a particular condition is suspected based on physical mutations and symptoms. Diagnostic testing can be performed at any time during a person's life, but is not available for all genes or all genetic conditions. The results of a diagnostic test can influence a person's choices about health care and the management of the disease. For example, people with a family history of polycystic kidney disease (PKD) who experience pain or tenderness in their abdomen, blood in their urine, frequent urination, pain in the sides, a urinary tract infection or kidney stones may decide to have their genes tested and the result could confirm the diagnosis of PKD.

- Carrier testing: Carrier testing is used to identify people who carry one copy of a gene mutation that, when present in two copies, causes a genetic disorder. This type of testing is offered to individuals who have a family history of a genetic disorder and to people in ethnic groups with an increased risk of specific genetic conditions. If both parents are tested, the test can provide information about a couple's risk of having a child with a genetic condition like cystic fibrosis.

- Preimplantation genetic diagnosis: Genetic testing procedures that are performed on human embryos prior to the implantation as part of an in vitro fertilization procedure. Pre-implantation testing is used when individuals try to conceive a child through in vitro fertilization. Eggs from the woman and sperm from the man are removed and fertilized outside the body to create multiple embryos. The embryos are individually screened for abnormalities, and the ones without abnormalities are implanted in the uterus.

- Prenatal diagnosis: Used to detect changes in a fetus's genes or chromosomes before birth. This type of testing is offered to couples with an increased risk of having a baby with a genetic or chromosomal disorder. In some cases, prenatal testing can lessen a couple's uncertainty or help them decide whether to abort the pregnancy. It cannot identify all possible inherited disorders and birth defects, however. One method of performing a prenatal genetic test involves an amniocentesis, which removes a sample of fluid from the mother's amniotic sac 15 to 20 or more weeks into pregnancy. The fluid is then tested for chromosomal abnormalities such as Down syndrome (Trisomy 21) and Trisomy 18, which can result in neonatal or fetal death. Test results can be retrieved within 7–14 days after the test is done. This method is 99.4% accurate at detecting and diagnosing fetal chromosome abnormalities. Although there is a risk of miscarriage associated with an amniocentesis, the miscarriage rate is only 1/400. Another method of prenatal testing is Chorionic Villus Sampling (CVS). Chorionic villi are projections from the placenta that carry the same genetic makeup as the baby. During this method of prenatal testing, a sample of chorionic villi is removed from the placenta to be tested. This test is performed 10–13 weeks into pregnancy and results are ready 7–14 days after the test was done. Another test using blood taken from the fetal umbilical cord is percutaneous umbilical cord blood sampling.

- Predictive and presymptomatic testing: Predictive and presymptomatic types of testing are used to detect gene mutations associated with disorders that appear after birth, often later in life. These tests can be helpful to people who have a family member with a genetic disorder, but who have no features of the disorder themselves at the time of testing. Predictive testing can identify mutations that increase a person's chances of developing disorders with a genetic basis, such as certain types of cancer. For example, an individual with a mutation in *BRCA1* has a 65% cumulative risk of breast cancer. Hereditary breast cancer along with ovarian cancer syndrome are caused by gene alterations in the genes BRCA1 and BRCA2. Major cancer types related to mutations in these genes are female breast cancer, ovarian, prostate, pancreatic, and male breast cancer. Li-Fraumeni syndrome is caused by a gene alteration on the gene TP53. Cancer types associated with a mutation on this gene include breast cancer, soft tissue sarcoma, osteosarcoma (bone cancer), leukemia and brain tumors. In the Cowden syndrome there is a mutation on the PTEN gene, causing potential breast, thyroid or endometrial cancer. Presymptomatic testing can determine whether a person will develop a genetic disorder, such as hemochromatosis (an iron overload disorder), before any signs or symptoms appear. The results of predictive and presymptomatic testing can provide information about a person's risk of developing a specific disorder, help with making decisions about medical care and provide a better prognosis.

- Pharmacogenomics: type of genetic testing that determines the influence of genetic variation on drug response. When a person has a disease or health condition, pharmacogenomics can examine an individual's genetic makeup to determine what medicine and what dosage would be the safest and most beneficial to the patient. In the human population, there are approximately 11 million single nucleotide polymorphisms (SNPs) in people's genomes, making them the most common variations in the human genome. SNPs reveal information about an individual's response to certain drugs. This type of genetic testing can be used for cancer patients undergoing chemotherapy. A sample of the cancer tissue can be sent in for genetic analysis by a specialized lab. After analysis, information retrieved can identify mutations in the tumor which can be used to determine the best treatment option.

Non-diagnostic testing includes:

- Forensic testing: Forensic testing uses DNA sequences to identify an individual for legal purposes. Unlike the tests described above, forensic testing is not used to detect gene mutations associated with disease. This type of testing can identify crime or catastrophe victims, rule out or implicate a crime suspect, or establish biological relationships between people (for example, paternity).

- Paternity testing: This type of genetic test uses special DNA markers to identify the same or similar inheritance patterns between related individuals. Based on the fact that we all inherit half of our DNA from the father, and half from the mother, DNA scientists test individuals to find the match of DNA sequences at some highly differential markers to draw the conclusion of relatedness.

- Genealogical DNA test: To determine ancestry or ethnic heritage for genetic genealogy.

- Research testing: Research testing includes finding unknown genes, learning how genes work and advancing our understanding of genetic conditions. The results of testing done as part of a research study are usually not available to patients or their healthcare providers.

Specific Diseases

Many diseases have a genetic component with tests already available. This list is continuously changing with additions of new test availabilities. This list below is just a few of the thousands of tests available.

- African iron overload

Over-absorption of iron; accumulation of iron in vital organs (heart, liver, pancreas); organ damage; heart disease; cancer; liver disease; arthritis; diabetes; infertility; impotence

- Alpha-1 antitrypsin deficiency

Obstructive lung disease in adults; liver cirrhosis during childhood; when a newborn or infant has jaundice that lasts for an extended period of time (more than a week or two), an enlarged spleen, ascites (fluid accumulation in the abdominal cavity), pruritus (itching), and other signs of liver injury; persons under 40 years of age that develops wheezing, a chronic cough or bronchitis, is short of breath after exertion and/or shows other signs of emphysema (especially when the patient is not a smoker, has not been exposed to known lung irritants, and when the lung damage appears to be located low in the lungs); when you have a close relative with alpha-1 antitrypsin deficiency; when a patient has a decreased level of A1AT.

- Apolipoprotein E-associated

Elevation of both serum cholesterol and triglycerides; accelerated atherosclerosis, coronary heart disease; cutaneous xanthomas; peripheral vascular disease; diabetes mellitus, obesity or hypothyroidism

- Becker/Duchenne muscular dystrophy

Muscle weakness (rapidly progressive); frequent falls; difficulty with motor skills (running, hopping, jumping); progressive difficulty walking (ability to walk may be lost by age 12); fatigue; intellectual retardation (possible); skeletal deformities; chest and back (scoliosis); muscle deformities (contractures of heels, legs; pseudohypertrophy of calf muscles)

- Beta-thalassemia

Reduced synthesis of the hemoglobin-beta chain; microcytic hypochromic anemia

- Factor II

Venous thrombosis; certain arterial thrombotic conditions; patients with deep vein thrombosis, pulmonary embolism, cerebral vein thrombosis, and premature ischemic stroke and also of women with premature myocardial infarction; family history of early onset stroke, deep vein thrombosis, thromboembolism, pregnancy associated with thrombosis/embolism, hyperhomocystinemia, and multiple miscarriage. Individuals with the mutation are at increased risk of thrombosis in the setting of oral contraceptive use, trauma, and surgery

- Factor V Leiden

Venous thrombosis; pulmonary embolism; transient ischemic attack or premature stroke; peripheral vascular disease, particularly lower extremity; occlusive disease; cerebral vein thrombosis; multiple spontaneous abortions; intrauterine fetal demise

- Homocysteine

Venous thrombosis; increased plasma homocysteine levels

- PAI-1 gene mutation

Independent risk factor for coronary artery disease, ischemic stroke, venous thrombosis (including osteonecrosis)

- Breast, ovarian and prostate cancer

Uncontrolled division of cancer cells

- Crohn's disease

Inflammation confined to the colon; abdominal pain and bloody diarrhea; anal fistulae and peri-rectal abscesses can also occur

- Cystic fibrosis

Large amount of abnormally thick mucus in the lungs and intestines; leads to congestioni, pneumonia, diarrhea and poor growth

- Deafness (non-syndromic)

Congenital loss of hearing; -prelingual, non-syndromic deafness

- Familial hypercholesterolemia

Tendon xanthomas; elevated LDL cholesterol; premature heart disease

- Fanconi anaemia

Predisposition of acute myeloid leukemia; skeletal abnormalities; radial hypoplasia and vertebral defect and other physical abnormalities, bone marrow failure (pancytopenia), endocrine dysfunction, early onset osteopenia/osteoporosis and lipid abnormalities, spontaneous chromosomal breakage exacerbated by exposure to DNA cross-linking agents

- Fragile-X syndrome

Mental retardation or learning disabilities of unknown etiology; autism or autistic-like characteristics; women with premature menopause. Subtle dysmorphism, log face with prominent mandible and large ears, macroorchidism in postpubertal males, behavioral abnormalities, due to lack of FMR1 in areas such as the cerebral cortex, amygdala, hippocampus and cerebellum

- Friedreich's ataxia

Characterized by slowly progressive ataxia; typically associated with depressed tendon reflexes, dysarthria, Babinski responses, and loss of position and vibration senses

- Hereditary hemochromatosis

over-absorption of iron; accumulation of iron in vital organs (heart, liver, pancreas); organ damage; heart disease; cancer; liver disease; arthritis; diabetes; infertility; impotence

- Hereditary hemochromatosis (Indian)
- Hirschsprung's disease

Absence of ganglia in the gut

- Huntington disease

Progressive disorder of motor, cognitive, and psychiatric disturbances

- Lactose Intolerance

Hypolactasia; persistent diarrhea; abdominal cramps; bloating; nausea; flatus

- Multiple endocrine neoplasia

MEN2A (which affects 60% to 90% of MEN2 families):Medullary thyroid carcinoma; Pheochromocytoma (tumor of the adrenal glands); Parathyroid adenomas (benign [noncancerous] tumors) or hyperplasia (increased size) of the parathyroid gland; MEN2B (which affects 5% of MEN2 families); Medullary thyroid carcinoma; Pheochromocytoma; Mucosal neuromas (benign tumors of nerve tissue on the tongue and lips); Digestive problems; Muscle, joint, and spinal problems; Typical facial features; Familial medullary thyroid carcinoma (FMTC) (which affects 5% to 35% of MEN2 families):Medullary thyroid carcinoma only

- Myotonic muscular dystrophy

Affects skeletal and smooth muscle as well as the eye, heart, endocrine system, and central nervous system; clinical findings, which span a continuum from mild to severe, have been categorized into three somewhat overlapping phenotypes: mild, classic, and congenital

- Pseudocholinesterase deficiency

Pseudocholinesterase (also called butyrylcholinesterase or "BCHE") hydrolyzes a number of choline-based compounds including cocaine, heroin, procaine, and succinylcholine, mivacurium, and other fast-acting muscle relaxants. Mutations in the BCHE gene lead to deficiency in the amount or function of the protein, which in turn results in a delay in the metabolism of these compounds, which prolongs their effects. Succinylcholine is commonly used as an anaesthetic in surgical procedures, and a person with BCHE mutations may suffer prolonged paraylasis. Between 1 in 3200 and 1 in 5000 people carry BCHE mutations; they are most prevalent in Persian Jews and Alaska Natives. As of 2013 there are 9 genetic tests available

- Sickle cell anaemia

Variable degrees of hemolysis and intermittent episodes of vascular occlusion resulting in tissue ischemia and acute and chronic organ dysfunction; complications include anemia, jaundice, predisposition to aplastic crisis, sepsis, cholelithiasis, and delayed growth. Diagnosis suspected in infants or young children with painful swelling of the hands and feet, pallor, jaundice, pneumococcal sepsis or meningitis, severe anemia with splenic enlargement, or acute chest syndrome

- Tay–Sachs disease

Lipids accumulate in the brain; neurological dysfunction; progressive weakness and loss of motor skills; decreased social interaction, seizures, blindness, and total debilitation

- Variegate porphyria

Cutaneous photosensitivity; acute neurovisceral crises

Medical Procedure

Genetic testing is often done as part of a genetic consultation and as of mid-2008 there were more than 1,200 clinically applicable genetic tests available. Once a person decides to proceed with genetic testing, a medical geneticist, genetic counselor, primary care doctor, or specialist can order the test after obtaining informed consent.

Genetic tests are performed on a sample of blood, hair, skin, amniotic fluid (the fluid that surrounds a fetus during pregnancy), or other tissue. For example, a medical procedure called a buccal smear uses a small brush or cotton swab to collect a sample of cells from the inside surface of the cheek. Alternatively, a small amount of saline mouthwash may be swished in the mouth to collect the cells. The sample is sent to a laboratory where technicians look for specific changes in chromosomes, DNA, or proteins, depending on the suspected disorders, often using DNA sequencing. The laboratory reports the test results in writing to a person's doctor or genetic counselor.

Routine newborn screening tests are done on a small blood sample obtained by pricking the baby's heel with a lancet.

Risks and Limitations

The physical risks associated with most genetic tests are very small, particularly for those tests that require only a blood sample or buccal smear (a procedure that samples cells from the inside surface of the cheek). The procedures used for prenatal testing carry a small but non-negligible risk of losing the pregnancy (miscarriage) because they require a sample of amniotic fluid or tissue from around the fetus.

Many of the risks associated with genetic testing involve the emotional, social, or financial consequences of the test results. People may feel angry, depressed, anxious, or guilty about their results. The potential negative impact of genetic testing has led to an increasing recognition of a "right not to know". In some cases, genetic testing creates tension within a family because the results can reveal information about other family members in addition to the person who is tested. The possibility of genetic discrimination in employment or insurance is also a concern. Some individuals avoid genetic testing out of fear it will affect their ability to purchase insurance or find a job. Health insurers do not currently require applicants for coverage to undergo genetic testing, and when insurers encounter genetic information, it is subject to the same confidentiality protections as any other sensitive health information. In the United States, the use of genetic information is governed by the Genetic Information Nondiscrimination Act (GINA).

Genetic testing can provide only limited information about an inherited condition. The test often can't determine if a person will show symptoms of a disorder, how severe the symptoms will be, or whether the disorder will progress over time. Another major limitation is the lack of treatment strategies for many genetic disorders once they are diagnosed.

A genetics professional can explain in detail the benefits, risks, and limitations of a particular test. It is important that any person who is considering genetic testing understand and weigh these factors before making a decision.

Other risks include accidental findings—a discovery of some possible problem found while looking for something else. In 2013 the American College of Medical Genetics and Genomics (ACMG) that certain genes always be included any time a genomic sequencing was done, and that labs should report the results.

Direct-to-consumer Genetic Testing

Direct-to-consumer (DTC) genetic testing is a type of genetic test that is accessible directly to the consumer without having to go through a health care professional. Usually, to obtain a genetic test, health care professionals (such as doctors) acquire their patient's permission and then order the desired test. DTC genetic tests, however, allow consumers to bypass this process and order DNA tests themselves.

There is a variety of DTC tests, ranging from tests for breast cancer alleles to mutations linked to cystic fibrosis. Benefits of DTC testing are the accessibility of tests to consumers, promotion of proactive healthcare, and the privacy of genetic information. Possible additional risks of DTC testing are the lack of governmental regulation, the potential misinterpretation of genetic information, issues related to testing minors, privacy of data, and downstream expenses for the public health care system.

Controversy

DTC genetic testing has been controversial due to outspoken opposition within the medical community. Critics of DTC testing argue against the risks involved, the unregulated advertising and marketing claims, and the overall lack of governmental oversight.

DTC testing involves many of the same risks associated with any genetic test. One of the more obvious and dangerous of these is the possibility of misreading of test results. Without professional guidance, consumers can potentially misinterpret genetic information, causing them to be deluded about their personal health.

Some advertising for DTC genetic testing has been criticized as conveying an exaggerated and inaccurate message about the connection between genetic information and disease risk, utilizing emotions as a selling factor. An advertisement for a BRCA-predictive genetic test for breast cancer stated: "There is no stronger antidote for fear than information."

Ancestry.com, a company providing DTC DNA tests for genealogy purposes, has reportedly allowed the warrantless search of their database by police investigating a murder. The warrantless search led to a search warrant to force the gathering of a DNA sample from a New Orleans filmmaker; however he turned out not to be a match for the suspected killer.

Government Regulation in The United States

Currently, the U.S. has no strong federal regulation moderating the DTC market. Though there are several hundred tests available, only a handful are approved by the Food and Drug Administration (FDA); these are sold as at-home test kits, and are therefore considered "medical devices" over which the FDA may assert jurisdiction. Other types of DTC tests require customers to mail in DNA samples for testing; it is difficult for the FDA to exercise jurisdiction over these types of tests, because the actual testing is completed in the laboratories of providers. As of 2007, the FDA had not yet officially substantiated with scientific evidence the claimed accuracy of the majority of direct-to-consumer genetic tests.

With regard to genetic testing and information in general, legislation in the United States called the Genetic Information Nondiscrimination Act prohibits group health plans and health insurers from denying coverage to a healthy individual or charging that person higher premiums based solely on a genetic predisposition to developing a disease in the future. The legislation also bars employers from using individuals' genetic information when making hiring, firing, job placement, or promotion decisions. The legislation, the first of its kind in the U.S., was passed by the United States Senate on April 24, 2008, on a vote of 95-0, and was signed into law by President George W. Bush on May 21, 2008. It went into effect on November 21, 2009.

In June 2013 the US Supreme Court issued two rulings on human genetics. The Court struck down patents on human genes, opening up competition in the field of genetic testing. The Supreme Court also ruled that police were allowed to collect DNA from people arrested for serious offenses.

In Popular Culture

Some possible future ethical problems of genetic testing were considered in the science fiction film *Gattaca*, the novel *Next*, and the science fiction anime series "Gundam Seed". Also, some films which include the topic of genetic testing include *The Island, Halloween: The Curse of Michael Myers*, and the *Resident Evil* series.

Ethics

Pediatric Genetic Testing

The American Academy of Pediatrics (AAP) and the American College of Medical Genetics (ACMG) have provided new guidelines for the ethical issue of pediatrics genetic testing and screening of children in the United States. Their guidelines state that performing pediatric genetic testing should be in the best interest of the child. In hypothetical situations for adults getting genetically tested 84-98% expressing interest in getting genetically tested for cancer predisposition. Though only half who are at risk of would get tested. AAP and ACMG recommend holding off on genetic testing for late-onset conditions until adulthood. Unless diagnosing genetic disorders during childhood and start early intervention can reduce morbidity or mortality. They also state that with parents or guardians permission testing for asymptomatic children who are at risk of childhood onset conditions are ideal reasons for pediatrics genetic testing. Testing for pharmacogenetics and newborn screening is found to be acceptable by AAP and ACMG guidelines. Histocompatibility testing guideline states that it's permissible for children of all ages to have tissue compatibility testing for immediate family members but only after the psychosocial, emotional and physical implications has been explored. With a donor advocate or similar mechanism should be in place to protect the minors from coercion and to safeguard the interest of said minor. Both AAP and ACMG discourage the use of direct-to-consumer and home kit genetic because of the accuracy, interpretation and oversight of test content. Guidelines also state that if parents or guardians should be encouraged to inform their child of the results from the genetic test if the minor is of appropriate age. If minor is of mature appropriate age and request results, the request should be honored. Though for ethical and legal reasons health care providers should be cautions in providing minors with predictive genetic testing without the involvement of parents or guardians. Within the guidelines AAP and ACMG state that health care provider have an obligation to inform parents or guardians on the implication of test results. To encourage patients and families to share information and even offer help in explain results to extend family or refer them to genetic counseling. AAP and ACMG state any type of predictive genetic testing for all types is best offer with genetic counseling being offer by Clinical genetics, genetic counselors or health care providers.

Israel

Israel uses DNA testing to determine if people are eligible for legal privileges given to specific ethnic groups. The policy where "many Jews from the Former Soviet Union ('FSU') are asked to provide DNA confirmation of their Jewish heritage in the form of paternety tests in order to immigrate as Jews and become citizens under Israel's Law of Return" has generated controversy.

Costs

The cost of genetic testing can range from under $100 to more than $2,000. This depends on the complexity of the test. The cost will increase if more than one test is necessary or if multiple family members are getting tested to obtain additional results. Costs can vary by state and some states cover part of the total cost.

From the date that a sample is taken, results may take weeks to months, depending upon the complexity and extent of the tests being performed. Results for prenatal testing are usually available more quickly because time is an important consideration in making decisions about a pregnancy. Prior to the testing, the doctor or genetic counselor who is requesting a particular test can provide specific information about the cost and time frame associated with that test.

Gene Delivery

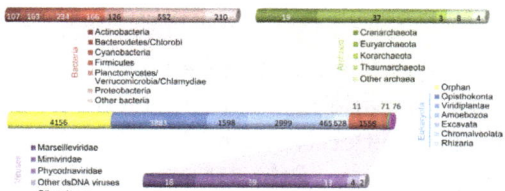

The eukaryotic *Acanthamoeba castellanii* genome organized by similarity to the four taxonomic kingdoms. This organism is noted for its high gene transfer rate, and is an example of gene delivery as a way to facilitate evolution.

Gene delivery is the process of introducing foreign DNA into host cells. Gene delivery occurs throughout nature as horizontal gene transfer from one organism to another and is a mechanism of evolution. It is also one of the steps necessary for gene therapy and, for example, has applications in the genetic modification of crops. There are many different methods of gene delivery for various types of cells and tissues, from bacterial to mammalian.

For gene delivery to be successful, foreign DNA must survive long enough in the host cell to integrate into its genome. This requires foreign DNA to be synthesized as part of a vector, which is designed to enter the desired host cell and deliver the transgene to that cell's genome. Vectors utilized as the method for gene delivery can be divided into two categories, non-viral and viral.

In complex multicellular eukaryotes (more specifically Weissmanists), if the transgene is incorporated into the host's germline cells, the resulting host cell can pass the transgene to its progeny. If the transgene is incorporated into somatic cells, the transgene will die with the somatic cell line, and thus its host organism.

Methods

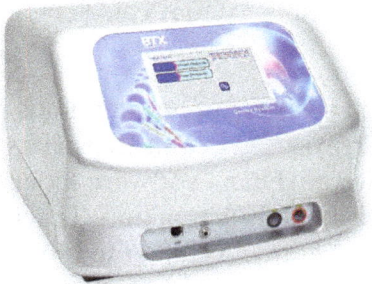

Electroporator with square wave and exponential decay waveforms for in vitro, in vivo, adherent cell and 96 well electroporation applications. Manufactured by BTX Harvard Apparatus, Holliston MA USA.

Non-Viral

Non-viral methods of gene delivery can be divided into transformation, where a cell incorporates foreign DNA from its surroundings, and conjugation, where gene transfer is a result of direct contact between two cells. Both methods utilize plasmids, which carry DNA inside a cell that can replicate independently of chromosomal DNA. This is the DNA that can be transferred to another organism. Artificial non-viral gene delivery can be mediated by physical methods such as electroporation, microinjection, gene gun, impalefection, hydrostatic pressure, continuous infusion, sonication and lipofection. It can also include the use of polymeric gene carriers (polyplexes).

Viral

Virus mediated gene delivery utilizes the ability of a virus to inject its DNA inside a host cell. Transduction is the process through which DNA is injected into the host cell and inserted into its genome. Viruses are a particularly effective form of gene delivery because the structure of the virus prevents degradation via lysosomes of the DNA it is delivering to the nucleus of the host cell. In gene therapy a gene that is intended for delivery is packaged into a replication-deficient viral particle to form a viral vector. Viruses used for gene therapy to date include retrovirus, adenovirus, adeno-associated virus and herpes simplex virus. However, there are drawbacks to using viruses to deliver genes into cells. Viruses can only deliver very small pieces of DNA into the cells, it is labor-intensive and there are risks of random insertion sites, cytophathic effects and mutagenesis.

Applications

In Nature

Gene delivery is an integral part of genome evolution. Genome analysis of *Acanthamoeba castellanii,* an amoebozoan, revealed that a significant portion of the genome that is expressed can be traced to lateral gene transfer amongst organisms. The ability of this eukaryote to host a wide range of endosymbionts allows it to interact with a wide range of DNA. When analyzed, the genome of *Acanthamoeba castellanii* shows genes with orthologs currently found in a diverse range of bacteria, viruses, and archaea.

Several of the methods used to facilitate gene delivery in nature and research have applications for therapeutic purposes. Gene therapy utilizes gene delivery to deliver genetic material with the goal of treating a disease or condition in the cell. Gene delivery in therapeutic settings utilizes non-immunogenic vectors capable of cell specificity that can deliver an adequate amount of transgene expression to cause the desired effect.

Advances in genomics have enabled a variety of new methods and gene targets to be identified for possible applications. DNA microarrays used in a variety of next-gen sequencing can identify thousands of genes simultaneously, with analytical software looking at gene expression patterns, and orthologous genes in model species to identify function. This has allowed a variety of possible vectors to be identified for use in gene therapy. As a method for creating a new class of vaccine, gene delivery has been utilized to generate a hybrid biosynthetic vector to deliver a possible vaccine. This vector overcomes traditional barriers to gene delivery by combining E. Coli with a synthetic polymer to create a vector that maintains plasmid DNA while having an increased ability to avoid degradation by target cell lysosomes.

Gene Therapy

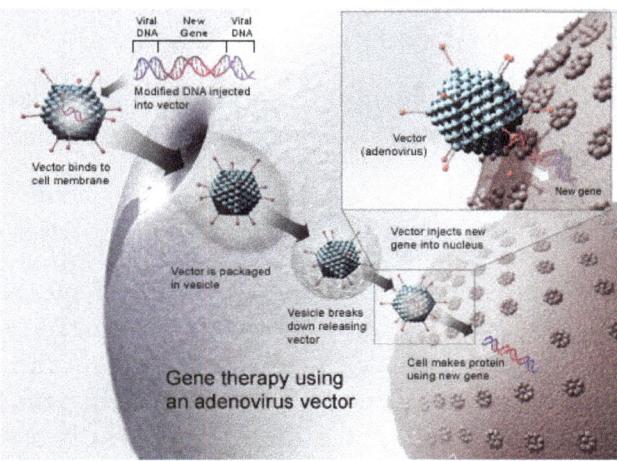

Gene therapy using an adenovirus vector. In some cases, the adenovirus will insert the new gene into a cell. If the treatment is successful, the new gene will make a functional protein to treat a disease.

Gene therapy is the therapeutic delivery of nucleic acid polymers into a patient's cells as a drug to treat disease. The first attempt at modifying human DNA was performed in 1980 by Martin Cline, but the first successful nuclear gene transfer in humans, approved by the National Institutes of Health, was performed in May 1989. The first therapeutic use of gene transfer as well as the first direct insertion of human DNA into the nuclear genome was performed by French Anderson in a trial starting in September 1990.

Between 1989 and February 2016, over 2,300 clinical trials had been conducted, more than half of them in phase I.

It should be noted that not all medical procedures that introduce alterations to a patient's genetic makeup can be considered gene therapy. Bone marrow transplantation and organ transplants in general have been found to introduce foreign DNA into patients. Gene therapy is defined by the precision of the procedure and the intention of direct therapeutic effects.

Background

Gene therapy was conceptualized in 1972, by authors who urged caution before commencing human gene therapy studies.

The first attempt, an unsuccessful one, at gene therapy (as well as the first case of medical transfer of foreign genes into humans not counting organ transplantation) was performed by Martin Cline on 10 July 1980. Cline claimed that one of the genes in his patients was active six months later, though he never published this data or had it verified and even if he is correct, it's unlikely it produced any significant beneficial effects treating beta-thalassemia.

After extensive research on animals throughout the 1980s and a 1989 bacterial gene tagging trial on humans, the first gene therapy widely accepted as a success was demonstrated in a trial that started on 14 September 1990, when Ashi DeSilva was treated for ADA-SCID.

The first somatic treatment that produced a permanent genetic change was performed in 1993.

This procedure was referred to sensationally and somewhat inaccurately in the media as a "three parent baby", though mtDNA is not the primary human genome and has little effect on an organism's individual characteristics beyond powering their cells.

Gene therapy is a way to fix a genetic problem at its source. The polymers are either translated into proteins, interfere with target gene expression, or possibly correct genetic mutations.

The most common form uses DNA that encodes a functional, therapeutic gene to replace a mutated gene. The polymer molecule is packaged within a "vector", which carries the molecule inside cells.

Early clinical failures led to dismissals of gene therapy. Clinical successes since 2006 regained researchers' attention, although as of 2014, it was still largely an experimental technique. These include treatment of retinal diseases Leber's congenital amaurosis and choroideremia, X-linked SCID, ADA-SCID, adrenoleukodystrophy, chronic lymphocytic leukemia (CLL), acute lymphocytic leukemia (ALL), multiple myeloma, haemophilia and Parkinson's disease. Between 2013 and April 2014, US companies invested over $600 million in the field.

The first commercial gene therapy, Gendicine, was approved in China in 2003 for the treatment of certain cancers. In 2011 Neovasculgen was registered in Russia as the first-in-class gene-therapy drug for treatment of peripheral artery disease, including critical limb ischemia. In 2012 Glybera, a treatment for a rare inherited disorder, became the first treatment to be approved for clinical use in either Europe or the United States after its endorsement by the European Commission.

Approaches

Following early advances in genetic engineering of bacteria, cells, and small animals, scientists started considering how to apply it to medicine. Two main approaches were considered – replacing or disrupting defective genes. Scientists focused on diseases caused by single-gene defects, such as cystic fibrosis, haemophilia, muscular dystrophy, thalassemia and sickle cell anemia. Glybera treats one such disease, caused by a defect in lipoprotein lipase.

DNA must be administered, reach the damaged cells, enter the cell and either express or disrupt a protein. Multiple delivery techniques have been explored. The initial approach incorporated DNA into an engineered virus to deliver the DNA into a chromosome. Naked DNA approaches have also been explored, especially in the context of vaccine development.

Generally, efforts focused on administering a gene that causes a needed protein to be expressed. More recently, increased understanding of nuclease function has led to more direct DNA editing, using techniques such as zinc finger nucleases and CRISPR. The vector incorporates genes into chromosomes. The expressed nucleases then knock out and replace genes in the chromosome. As of 2014 these approaches involve removing cells from patients, editing a chromosome and returning the transformed cells to patients.

Gene editing is a potential approach to alter the human genome to treat genetic diseases, viral diseases, and cancer. As of 2016 these approaches were still years from being medicine.

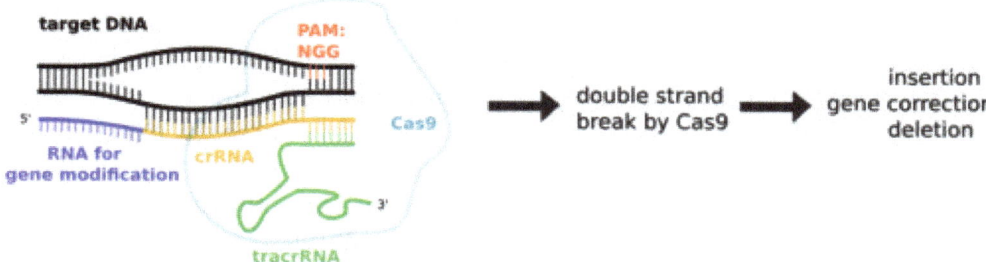

A duplex of crRNA and tracrRNA acts as guide RNA to introduce a specifically located gene modification based on the RNA 5' upstream of the crRNA. Cas9 binds the tracrRNA and needs a DNA binding sequence (5'NGG3'), which is called protospacer adjacent motif (PAM). After binding, Cas9 introduces a DNA double strand break, which is then followed by gene modification via homologous recombination (HDR) or non-homologous end joining (NHEJ).

Cell Types

Gene therapy may be classified into two types:

Somatic

In somatic cell gene therapy (SCGT), the therapeutic genes are transferred into any cell other than a gamete, germ cell, gametocyte or undifferentiated stem cell. Any such modifications affect the individual patient only, and are not inherited by offspring. Somatic gene therapy represents mainstream basic and clinical research, in which therapeutic DNA (either integrated in the genome or as an external episome or plasmid) is used to treat disease.

Over 600 clinical trials utilizing SCGT are underway in the US. Most focus on severe genetic disorders, including immunodeficiencies, haemophilia, thalassaemia and cystic fibrosis. Such single gene disorders are good candidates for somatic cell therapy. The complete correction of a genetic disorder or the replacement of multiple genes is not yet possible. Only a few of the trials are in the advanced stages.

Germline

In germline gene therapy (GGT), germ cells (sperm or eggs) are modified by the introduction of functional genes into their genomes. Modifying a germ cell causes all the organism's cells to contain the modified gene. The change is therefore heritable and passed on to later generations. Australia, Canada, Germany, Israel, Switzerland and the Netherlands prohibit GGT for application in human beings, for technical and ethical reasons, including insufficient knowledge about possible risks to future generations and higher risks versus SCGT. The US has no federal controls specifically addressing human genetic modification (beyond FDA regulations for therapies in general).

Vectors

The delivery of DNA into cells can be accomplished by multiple methods. The two major classes are recombinant viruses (sometimes called biological nanoparticles or viral vectors) and naked DNA or DNA complexes (non-viral methods).

Viruses

In order to replicate, viruses introduce their genetic material into the host cell, tricking the host's cellular machinery into using it as blueprints for viral proteins. Scientists exploit this by substituting a virus's genetic material with therapeutic DNA. (The term 'DNA' may be an oversimplification, as some viruses contain RNA, and gene therapy could take this form as well.) A number of viruses have been used for human gene therapy, including retrovirus, adenovirus, lentivirus, herpes simplex, vaccinia and adeno-associated virus. Like the genetic material (DNA or RNA) in viruses, therapeutic DNA can be designed to simply serve as a temporary blueprint that is degraded naturally or (at least theoretically) to enter the host's genome, becoming a permanent part of the host's DNA in infected cells.

Non-viral

Non-viral methods present certain advantages over viral methods, such as large scale production and low host immunogenicity. However, non-viral methods initially produced lower levels of transfection and gene expression, and thus lower therapeutic efficacy. Later technology remedied this deficiency.

Methods for non-viral gene therapy include the injection of naked DNA, electroporation, the gene gun, sonoporation, magnetofection, the use of oligonucleotides, lipoplexes, dendrimers, and inorganic nanoparticles.

Hurdles

Some of the unsolved problems include:

- Short-lived nature – Before gene therapy can become a permanent cure for a condition, the therapeutic DNA introduced into target cells must remain functional and the cells containing the therapeutic DNA must be stable. Problems with integrating therapeutic DNA into the genome and the rapidly dividing nature of many cells prevent it from achieving long-term benefits. Patients require multiple treatments.

- Immune response – Any time a foreign object is introduced into human tissues, the immune system is stimulated to attack the invader. Stimulating the immune system in a way that reduces gene therapy effectiveness is possible. The immune system's enhanced response to viruses that it has seen before reduces the effectiveness to repeated treatments.

- Problems with viral vectors – Viral vectors carry the risks of toxicity, inflammatory responses, and gene control and targeting issues.

- Multigene disorders – Some commonly occurring disorders, such as heart disease, high blood pressure, Alzheimer's disease, arthritis, and diabetes, are affected by variations in multiple genes, which complicate gene therapy.

- Some therapies may breach the Weismann barrier (between soma and germ-line) protecting the testes, potentially modifying the germline, falling afoul of regulations in countries that prohibit the latter practice.

- Insertional mutagenesis – If the DNA is integrated in a sensitive spot in the genome, for example in a tumor suppressor gene, the therapy could induce a tumor. This has occurred in clinical trials for X-linked severe combined immunodeficiency (X-SCID) patients, in which hematopoietic stem cells were transduced with a corrective transgene using a retrovirus, and this led to the development of T cell leukemia in 3 of 20 patients. One possible solution is to add a functional tumor suppressor gene to the DNA to be integrated. This may be problematic since the longer the DNA is, the harder it is to integrate into cell genomes. CRISPR technology allows researchers to make much more precise genome changes at exact locations.

- Cost – Alipogene tiparvovec or Glybera, for example, at a cost of $1.6 million per patient, was reported in 2013 to be the world's most expensive drug.

Deaths

Three patients' deaths have been reported in gene therapy trials, putting the field under close scrutiny. The first was that of Jesse Gelsinger in 1999. One X-SCID patient died of leukemia in 2003. In 2007, a rheumatoid arthritis patient died from an infection; the subsequent investigation concluded that the death was not related to gene therapy.

History

1970s and Earlier

In 1972 Friedmann and Roblin authored a paper in *Science* titled "Gene therapy for human genetic disease?" Rogers (1970) was cited for proposing that *exogenous good DNA* be used to replace the defective DNA in those who suffer from genetic defects.

1980s

In 1984 a retrovirus vector system was designed that could efficiently insert foreign genes into mammalian chromosomes.

1990s

The first approved gene therapy clinical research in the US took place on 14 September 1990, at the National Institutes of Health (NIH), under the direction of William French Anderson. Four-year-old Ashanti DeSilva received treatment for a genetic defect that left her with ADA-SCID, a severe immune system deficiency. The effects were temporary, but successful.

Cancer gene therapy was introduced in 1992/93 (Trojan et al. 1993). The treatment of glioblastoma multiforme, the malignant brain tumor whose outcome is always fatal, was done using a vector expressing antisense IGF-I RNA (clinical trial approved by NIH protocolno.1602 November 24, 1993, and by the FDA in 1994). This therapy also represents the beginning of cancer immunogene therapy, a treatment which proves to be effective due to the anti-tumor mechanism of IGF-I antisense, which is related to strong immune and apoptotic phenomena.

In 1992 Claudio Bordignon, working at the Vita-Salute San Raffaele University, performed the first gene therapy procedure using hematopoietic stem cells as vectors to deliver genes intended to correct hereditary diseases. In 2002 this work led to the publication of the first success-ful gene therapy treatment for adenosine deaminase deficiency (ADA-SCID). The success of a multi-center trial for treating children with SCID (severe combined immune deficiency or "bub-ble boy" disease) from 2000 and 2002, was questioned when two of the ten children treated at the trial's Paris center developed a leukemia-like condition. Clinical trials were halted temporar-ily in 2002, but resumed after regulatory review of the protocol in the US, the United Kingdom, France, Italy and Germany.

In 1993 Andrew Gobea was born with SCID following prenatal genetic screening. Blood was re-moved from his mother's placenta and umbilical cord immediately after birth, to acquire stem cells. The allele that codes for adenosine deaminase (ADA) was obtained and inserted into a ret-rovirus. Retroviruses and stem cells were mixed, after which the viruses inserted the gene into the stem cell chromosomes. Stem cells containing the working ADA gene were injected into Andrew's blood. Injections of the ADA enzyme were also given weekly. For four years T cells (white blood cells), produced by stem cells, made ADA enzymes using the ADA gene. After four years more treatment was needed.

Jesse Gelsinger's death in 1999 impeded gene therapy research in the US. As a result, the FDA suspended several clinical trials pending the reevaluation of ethical and procedural practices.

2000s

The modified cancer gene therapy strategy of antisense IGF-I RNA (NIH n° 1602) using antisense / triple helix anti IGF-I approach was registered in 2002 by Wiley gene therapy clinical trial - n° 635 and 636. The approach has shown promising results in the treatment of six different malig-nant tumors: glioblastoma, cancers of liver, colon, prostate, uterus and ovary (Collaborative NATO Science Programme on Gene Therapy USA, France, Poland n° LST 980517 conducted by J. Trojan) (Trojan et al., 2012). This anti–gene antisense/triple helix therapy has proven to be efficient, due to the mechanism stopping simultaneously IGF-I expression on translation and transcription lev-els, strengthening anti-tumor immune and apoptotic phenomena.

2002

Sickle-cell disease can be treated in mice. The mice – which have essentially the same defect that causes human cases – used a viral vector to induce production of fetal hemoglobin (HbF), which normally ceases to be produced shortly after birth. In humans, the use of hydroxyurea to stimulate the production of HbF temporarily alleviates sickle cell symptoms. The researchers demonstrated this treatment to be a more permanent means to increase therapeutic HbF production.

A new gene therapy approach repaired errors in messenger RNA derived from defective genes. This technique has the potential to treat thalassaemia, cystic fibrosis and some cancers.

Researchers created liposomes 25 nanometers across that can carry therapeutic DNA through pores in the nuclear membrane.

2003

In 2003 a research team inserted genes into the brain for the first time. They used liposomes coated in a polymer called polyethylene glycol, which, unlike viral vectors, are small enough to cross the blood–brain barrier.

Short pieces of double-stranded RNA (short, interfering RNAs or siRNAs) are used by cells to degrade RNA of a particular sequence. If a siRNA is designed to match the RNA copied from a faulty gene, then the abnormal protein product of that gene will not be produced.

Gendicine is a cancer gene therapy that delivers the tumor suppressor gene p53 using an engineered adenovirus. In 2003, it was approved in China for the treatment of head and neck squamous cell carcinoma.

2006

In March researchers announced the successful use of gene therapy to treat two adult patients for X-linked chronic granulomatous disease, a disease which affects myeloid cells and damages the immune system. The study is the first to show that gene therapy can treat the myeloid system.

In May a team reported a way to prevent the immune system from rejecting a newly delivered gene. Similar to organ transplantation, gene therapy has been plagued by this problem. The immune system normally recognizes the new gene as foreign and rejects the cells carrying it. The research utilized a newly uncovered network of genes regulated by molecules known as microRNAs. This natural function selectively obscured their therapeutic gene in immune system cells and protected it from discovery. Mice infected with the gene containing an immune-cell microRNA target sequence did not reject the gene.

In August scientists successfully treated metastatic melanoma in two patients using killer T cells genetically retargeted to attack the cancer cells.

In November researchers reported on the use of VRX496, a gene-based immunotherapy for the treatment of HIV that uses a lentiviral vector to deliver an antisense gene against the HIV envelope. In a phase I clinical trial, five subjects with chronic HIV infection who had failed to respond to at least two antiretroviral regimens were treated. A single intravenous infusion of autologous CD4 T cells genetically modified with VRX496 was well tolerated. All patients had stable or decreased viral load; four of the five patients had stable or increased CD4 T cell counts. All five patients had stable or increased immune response to HIV antigens and other pathogens. This was the first evaluation of a lentiviral vector administered in a US human clinical trial.

2007

In May researchers announced the first gene therapy trial for inherited retinal disease. The first operation was carried out on a 23-year-old British male, Robert Johnson, in early 2007.

2008

Leber's congenital amaurosis is an inherited blinding disease caused by mutations in the RPE65 gene. The results of a small clinical trial in children were published in April. Delivery of recombinant adeno-associated virus (AAV) carrying RPE65 yielded positive results. In May two more groups reported positive results in independent clinical trials using gene therapy to treat the condition. In all three clinical trials, patients recovered functional vision without apparent side-effects.

2009

In September researchers were able to give trichromatic vision to squirrel monkeys. In November 2009, researchers halted a fatal genetic disorder called adrenoleukodystrophy in two children using a lentivirus vector to deliver a functioning version of ABCD1, the gene that is mutated in the disorder.

2010s

2010

An April paper reported that gene therapy addressed achromatopsia (color blindness) in dogs by targeting cone photoreceptors. Cone function and day vision were restored for at least 33 months in two young specimens. The therapy was less efficient for older dogs.

In September it was announced that an 18-year-old male patient in France with beta-thalassemia major had been successfully treated. Beta-thalassemia major is an inherited blood disease in which beta haemoglobin is missing and patients are dependent on regular lifelong blood transfusions. The technique used a lentiviral vector to transduce the human ß-globin gene into purified blood and marrow cells obtained from the patient in June 2007. The patient's haemoglobin levels were stable at 9 to 10 g/dL. About a third of the hemoglobin contained the form introduced by the viral vector and blood transfusions were not needed. Further clinical trials were planned. Bone marrow transplants are the only cure for thalassemia, but 75% of patients do not find a matching donor.

Cancer immunogene therapy using modified anti – gene, antisense / triple helix approach was introduced in South America in 2010/11 in La Sabana University, Bogota (Ethical Committee 14 December 2010, no P-004-10). Considering the ethical aspect of gene diagnostic and gene therapy targeting IGF-I, the IGF-I expressing tumors i.e. lung and epidermis cancers, were treated (Trojan et al. 2016).

2011

In 2007 and 2008, a man (Timothy Ray Brown) was cured of HIV by repeated hematopoietic stem cell transplantation with double-delta-32 mutation which disables the CCR5 receptor. This cure was accepted by the medical community in 2011. It required complete ablation of existing bone marrow, which is very debilitating.

In August two of three subjects of a pilot study were confirmed to have been cured from chronic lymphocytic leukemia (CLL). The therapy used genetically modified T cells to attack cells that expressed the CD19 protein to fight the disease. In 2013, the researchers announced that 26 of 59 patients had achieved complete remission and the original patient had remained tumor-free.

Human HGF plasmid DNA therapy of cardiomyocytes is being examined as a potential treatment for coronary artery disease as well as treatment for the damage that occurs to the heart after myocardial infarction.

In 2011 Neovasculgen was registered in Russia as the first-in-class gene-therapy drug for treatment of peripheral artery disease, including critical limb ischemia; it delivers the gene encoding for VEGF. Neovasculogen is a plasmid encoding the CMV promoter and the 165 amino acid form of VEGF.

2012

The FDA approved Phase 1 clinical trials on thalassemia major patients in the US for 10 participants in July. The study was expected to continue until 2015.

In July 2012, the European Medicines Agency recommended approval of a gene therapy treatment for the first time in either Europe or the United States. The treatment used Alipogene tiparvovec (Glybera) to compensate for lipoprotein lipase deficiency, which can cause severe pancreatitis. The recommendation was endorsed by the European Commission in November 2012 and commercial rollout began in late 2014. Alipogene tiparvovec was expected to cost around $1.6 million per treatment in 2012, revised to $1 million in 2015, making it the most expensive medicine in the world at the time. As of 2016, only one person had been treated with drug.

In December 2012, it was reported that 10 of 13 patients with multiple myeloma were in remission "or very close to it" three months after being injected with a treatment involving genetically engineered T cells to target proteins NY-ESO-1 and LAGE-1, which exist only on cancerous myeloma cells.

2013

In March researchers reported that three of five adult subjects who had acute lymphocytic leukemia (ALL) had been in remission for five months to two years after being treated with genetically modified T cells which attacked cells with CD19 genes on their surface, i.e. all B-cells, cancerous or not. The researchers believed that the patients' immune systems would make normal T-cells and B-cells after a couple of months. They were also given bone marrow. One patient relapsed and died and one died of a blood clot unrelated to the disease.

Following encouraging Phase 1 trials, in April, researchers announced they were starting Phase 2 clinical trials (called CUPID2 and SERCA-LVAD) on 250 patients at several hospitals to combat heart disease. The therapy was designed to increase the levels of SERCA2, a protein in heart muscles, improving muscle function. The FDA granted this a Breakthrough Therapy Designation to accelerate the trial and approval process. In 2016 it was reported that no improvement was found from the CUPID 2 trial.

In July researchers reported promising results for six children with two severe hereditary diseases had been treated with a partially deactivated lentivirus to replace a faulty gene and after 7–32 months. Three of the children had metachromatic leukodystrophy, which causes children to lose cognitive and motor skills. The other children had Wiskott-Aldrich syndrome, which leaves them to open to infection, autoimmune diseases and cancer. Follow up trials with gene therapy on another six children with Wiskott-Aldrich syndrome were also reported as promising.

In October researchers reported that two children born with adenosine deaminase severe combined immunodeficiency disease (ADA-SCID) had been treated with genetically engineered stem cells 18 months previously and that their immune systems were showing signs of full recovery. Another three children were making progress. In 2014 a further 18 children with ADA-SCID were cured by gene therapy. ADA-SCID children have no functioning immune system and are sometimes known as "bubble children."

Also in October researchers reported that they had treated six haemophilia sufferers in early 2011 using an adeno-associated virus. Over two years later all six were producing clotting factor.

2014

In January researchers reported that six choroideremia patients had been treated with adeno-associated virus with a copy of REP1. Over a six-month to two-year period all had improved their sight. By 2016, 32 patients had been treated with positive results and researchers were hopeful the treatment would be long-lasting. Choroideremia is an inherited genetic eye disease with no approved treatment, leading to loss of sight.

In March researchers reported that 12 HIV patients had been treated since 2009 in a trial with a genetically engineered virus with a rare mutation (CCR5 deficiency) known to protect against HIV with promising results.

Clinical trials of gene therapy for sickle cell disease were started in 2014. There is a need for high quality randomised controlled trials assessing the risks and benefits invoved with gene therapy for people with sickle cell disease.

2015

In February LentiGlobin BB305, a gene therapy treatment undergoing clinical trials for treatment of beta thalassemia gained FDA "breakthrough" status after several patients were able to forgo the frequent blood transfusions usually required to treat the disease.

In March researchers delivered a recombinant gene encoding a broadly neutralizing antibody into monkeys infected with simian HIV; the monkeys' cells produced the antibody, which cleared them of HIV. The technique is named immunoprophylaxis by gene transfer (IGT). Animal tests for antibodies to ebola, malaria, influenza and hepatitis were underway.

In March, scientists, including an inventor of CRISPR, urged a worldwide moratorium on germline gene therapy, writing "scientists should avoid even attempting, in lax jurisdictions, germline genome modification for clinical application in humans" until the full implications "are discussed among scientific and governmental organizations".

In October, researchers announced that they had treated a baby girl, Layla Richards, with an experimental treatment using donor T-cells genetically engineered using TALEN to attack cancer cells. One year after the treatment she was still free of her cancer (a highly aggressive form of acute lymphoblastic leukaemia [ALL]). Children with highly aggressive ALL normally have a very poor prognosis and Layla's disease had been regarded as terminal before the treatment.

In December, scientists of major world academies called for a moratorium on inheritable human genome edits, including those related to CRISPR-Cas9 technologies but that basic research including embryo gene editing should continue.

2016

In April the Committee for Medicinal Products for Human Use of the European Medicines Agency endorsed a gene therapy treatment called Strimvelis and the European Commission approved it in June. This treats children born with ADA-SCID and who have no functioning immune system - sometimes called the "bubble baby" disease. This was the second gene therapy treatment to be approved in Europe.

In October, Chinese scientists reported they had started a trial to genetically modify T-cells from 10 adult patients with lung cancer and reinject the modified T-cells back into their bodies to attack the cancer cells. The T-cells had the PD-1 protein (which stops or slows the immune response) removed using CRISPR-Cas9.

A 2016 Cochrane systematic review looking at data from four trials on topical cystic fibrosis transmembrane conductance regulator (CFTR) gene therapy does not support its clinical use as a mist inhaled into the lungs to treat cystic fibrosis patients with lung infections. One of the four trials did find weak evidence that liposome-based CFTR gene transfer therapy may lead to a small respiratory improvement for people with CF. This weak evidence is not enough to make a clinical recommendation for routine CFTR gene therapy.

2017

In February Kite Pharma announced results from a clinical trial of CAR-T cells in around a hundred people with advanced Non-Hodgkin lymphoma.

In March, French scientists reported on clinical research of gene therapy to treat sickle-cell disease.

Speculative Uses

Speculated uses for gene therapy include:

Fertility

Gene Therapy techniques have the potential to provide alternative treatments for those with infertility. Recently, successful experimentation on mice has proven that fertility can be restored by using the gene therapy method, CRISPR. Spermatogenical stem cells from another organism were transplanted into the testes of an infertile male mouse. The stem cells re-established spermatogenesis and fertility.

Gene Doping

Athletes might adopt gene therapy technologies to improve their performance. Gene doping is not known to occur, but multiple gene therapies may have such effects. Kayser et al. argue that gene doping could level the playing field if all athletes receive equal access. Critics claim that any therapeutic intervention for non-therapeutic/enhancement purposes compromises the ethical foundations of medicine and sports.

Human Genetic Engineering

Genetic engineering could be used to change physical appearance, metabolism, and even improve physical capabilities and mental faculties such as memory and intelligence. Ethical claims about germline engineering include beliefs that every fetus has a right to remain genetically unmodified, that parents hold the right to genetically modify their offspring, and that every child has the right to be born free of preventable diseases. For parents, genetic engineering could be seen as another child enhancement technique to add to diet, exercise, education, training, cosmetics and plastic surgery. Another theorist claims that moral concerns limit but do not prohibit germline engineering.

Possible regulatory schemes include a complete ban, provision to everyone, or professional self-regulation. The American Medical Association's Council on Ethical and Judicial Affairs stated that "genetic interventions to enhance traits should be considered permissible only in severely restricted situations: (1) clear and meaningful benefits to the fetus or child; (2) no trade-off with other characteristics or traits; and (3) equal access to the genetic technology, irrespective of income or other socioeconomic characteristics."

As early in the history of biotechnology as 1990, there have been scientists opposed to attempts to modify the human germline using these new tools, and such concerns have continued as technology progressed. With the advent of new techniques like CRISPR, in March 2015 a group of scientists urged a worldwide moratorium on clinical use of gene editing technologies to edit the human genome in a way that can be inherited. In April 2015, researchers sparked controversy when they reported results of basic research to edit the DNA of non-viable human embryos using CRISPR. A committee of the American National Academy of Sciences and National Academy of Medicine gave qualified support to human genome editing in 2017 once answers have been found to safety and efficiency problems "but only for serious conditions under stringent oversight."

Regulations

Regulations covering genetic modification are part of general guidelines about human-involved biomedical research.

The Helsinki Declaration (Ethical Principles for Medical Research Involving Human Subjects) was amended by the World Medical Association's General Assembly in 2008. This document provides principles physicians and researchers must consider when involving humans as research subjects. The Statement on Gene Therapy Research initiated by the Human Genome Organization (HUGO) in 2001 provides a legal baseline for all countries. HUGO's document emphasizes human freedom and adherence to human rights, and offers recommendations for somatic gene therapy, including the importance of recognizing public concerns about such research.

United States

No federal legislation lays out protocols or restrictions about human genetic engineering. This subject is governed by overlapping regulations from local and federal agencies, including the Department of Health and Human Services, the FDA and NIH's Recombinant DNA Advisory Committee. Researchers seeking federal funds for an investigational new drug application, (commonly the case for somatic human genetic engineering), must obey international and federal guidelines for the protection of human subjects.

NIH serves as the main gene therapy regulator for federally funded research. Privately funded research is advised to follow these regulations. NIH provides funding for research that develops or enhances genetic engineering techniques and to evaluate the ethics and quality in current research. The NIH maintains a mandatory registry of human genetic engineering research protocols that includes all federally funded projects.

An NIH advisory committee published a set of guidelines on gene manipulation. The guidelines discuss lab safety as well as human test subjects and various experimental types that involve genetic changes. Several sections specifically pertain to human genetic engineering, including Section III-C-1. This section describes required review processes and other aspects when seeking approval to begin clinical research involving genetic transfer into a human patient. The protocol for a gene therapy clinical trial must be approved by the NIH's Recombinant DNA Advisory Committee prior to any clinical trial beginning; this is different from any other kind of clinical trial.

As with other kinds of drugs, the FDA regulates the quality and safety of gene therapy products and supervises how these products are used clinically. Therapeutic alteration of the human genome falls under the same regulatory requirements as any other medical treatment. Research involving human subjects, such as clinical trials, must be reviewed and approved by the FDA and an Institutional Review Board.

Popular Culture

Gene therapy is the basis for the plotline of the film *I Am Legend* and the TV show *Will Gene Therapy Change the Human Race?*. It is also used in Stargate as a means of allowing humans to use ancient technology.

Gene Doping

Gene doping is the hypothetical non-therapeutic use of gene therapy in order to improve athletic performance in competitive sporting events. As of April 2015, there is no evidence that gene doping has been used for athletic performance-enhancement in any sporting events. Gene doping would involve the use of gene transfer to increase or decrease gene expression and protein biosynthesis of a specific human protein; this could be done by directly injecting the gene carrier into the person, or by taking cells from the person, transfecting the cells, and administering the cells back to the person.

The historical development of interest in gene doping by athletes and concern about the risks of gene doping and how to detect it moved in parallel with the development of the field of gene therapy, especially with the publication in 1998 of work on a transgenic mouse overexpressing insulin-like growth factor 1 that was much stronger than normal mice, even in old age, preclinical studies published in 2002 of a way to deliver erythropoietin (EPO) via gene therapy, and publication in 2004 of the creation of a "marathon mouse" with much greater endurance than normal mice, created by delivering the gene expressing PPAR gamma to the mice. The scientists generating these publications were all contacted directly by athletes and coaches seeking access to the technology. The public became aware of that activity in 2006 when such efforts were part of the evidence presented in the trial of a German coach.

Scientists themselves, as well as bodies including the World Anti-Doping Agency (WADA), the International Olympic Committee, and the American Association for the Advancement of Science, started discussing the risk of gene doping in 2001, and by 2003 WADA had added gene doping to the list of banned doping practices, and shortly thereafter began funding research on methods to detect gene doping.

History

The history of concern about the potential for gene doping follows the history of gene therapy, the medical use of genes to treat diseases, which was first clinically tested in the 1990s. Interest by the athletic community was especially spurred by the creation in a university lab of a "mighty mouse", created by administering a virus carrying the gene expressing insulin-like growth factor 1 to mice; the mice were stronger and remained strong even as they aged, without exercise. The lab had been seeking treatments for muscle wasting diseases, but when their work was made public, the lab was inundated with calls from athletes seeking the treatment, with one coach offering his whole team. The scientist told *The New York Times* in 2007: "I was quite surprised, I must admit. People would try to entice me, saying things like, 'It'll help advance your research.' Some offered to pay me." He also told the *Times* that every time similar research is published he gets calls and that he explains that, even should the treatment became ready for use in people, which would take years, there would be serious risks, including death; he also said that even after he explains this, the athletes still want it.

In 1999, the field of gene therapy was set back when Jesse Gelsinger died in a gene therapy clinical trial, suffering a massive inflammatory reaction to the drug. This led regulatory authorities in the US and Europe to increase safety requirements in clinical trials even beyond the initial restrictions that had been put in place at the beginning of the biotechnology era to deal with the risks of recombinant DNA.

In June 2001, Theodore Friedmann, one of the pioneers of gene therapy, and Johann Olav Koss an Olympic gold medallist in speed skating, published a paper that was the first public warning about gene doping. Also in June 2001, a Gene Therapy Working Group, convened by the Medical Commission of the International Olympic Committee noted that "we are aware that there is the potential for abuse of gene therapy medicines and we shall begin to establish procedures and state-of-the-art testing methods for identifying athletes who might misuse such technology".

Research was published in 2002 about a preclinical gene therapy called Repoxygen, which delivered the gene encoding erythropoietin (EPO) as a potential treatment for anemia. The scientists from that company also received calls from athletes and coaches. In that same year the World Anti-Doping Agency held its first meeting to discuss the risk of gene doping, and the US The President's Council on Bioethics discussed gene doping in the context of human enhancement at several sessions.

In 2003, the field of gene therapy took a step forward and a step back; first gene therapy drug was approved, Gendicine, which was approved in China for the treatment of certain cancers, but children in France who had seemingly been effective treated with gene therapy for severe combined immunodeficiency (non-human) began developing leukemia. In 2003 the BALCO scandal became public, in which chemists, trainers and athletes conspired to evade doping controls with new and undetectable doping substances. In 2003 the World Doping Agency proactively added gene doping to the list of banned doping practices. Also in 2003, a symposium convened by the American Association for the Advancement of Science focused on the issue.

Research published in 2004 showing that mice given gene therapy coding for a protein called PPAR gamma had about double the endurance of untreated mice and were dubbed "marathon mice"; those scientists received calls from athletes and coaches. Also in 2004 the World Anti-Doping Agency began to fund research to detect gene doping, and formed a permanent expert panel to advise it on risks and to guide the funding.

In 2006 interest from athletes in gene doping received widespread media coverage due its mention during the trial of a German coach who was accused and found guilty of giving his athletes performance enhancing drugs without their knowledge; an email in which the coach attempted to obtain Repoxygen was read in open court by a prosecutor. This was the first public disclosure that athletes were interested in gene doping.

In 2011 the second gene therapy drug was approved; Neovasculgen, which delivers the gene encoding VEGF, was approved in Russia to treat peripheral artery disease.

In 2012 Glybera, a treatment for a rare inherited disorder, became the first treatment to be approved for clinical use in either Europe or the United States.

As the field of gene therapy has developed, the risk of gene doping becoming a reality has increased with it.

Agents

There are numerous genes of interest as agents for gene doping. They include erythropoietin, insulin-like growth factor 1, human growth hormone, myostatin, vascular endothelial growth factor, fibroblast growth factor, endorphin, enkephalin and alpha-actinin-3.

The risks of gene doping would be similar to those of gene therapy: immune reaction to the native protein leading to the equivalent of an genetic disease, massive inflammatory response, cancer, and death, and in all cases, these risks would be undertaken for short-term gain as opposed to treating a serious disease.

Alpha-actinin-3

Alpha-actinin-3 is found only in skeletal muscle in humans, and has been identified in several genetic studies as having a different polymorphism in world-class athletes compared with normal people. One form that causes the gene to make more protein is found in sprinters and is related to increased power; another form that causes the gene to make less protein is found in endurance athletes. Gene doping agents could be designed with either polymorphism, or for endurance athletes, some DNA construct that interfered with expression like a small interfering RNA.

Myostatin

Myostatin is a protein responsible for inhibiting muscle differentiation and growth. Removing the myostatin gene or otherwise limiting its expression leads to an increase in muscle size and power. This has been demonstrated in knockout mice lacking the gene that were dubbed "Schwarzenegger mice". Humans born with defective genes can also serve as "knockout models"; a German boy with a mutation in both copies of the myostatin gene was born with well-developed muscles. The advanced muscle growth continued after birth, and the boy could lift weights of 3 kg at the age of 4. In work published in 2009, scientists administered follistatin via gene therapy to the quadriceps of non-human primates, resulting in local muscle growth similar to the mice.

Erythropoietin (EPO)

Erythropoietin is a glycoprotein that acts as a hormone, controlling red blood cell production. Athletes have injected the EPO protein as a performance-enhancing substance for many years (blood doping). When the additional EPO increases the production of red blood cells in circulation, this increases the amount of oxygen available to muscle, enhancing an athlete's endurance. Recent studies suggest it may be possible to introduce another EPO gene into an animal in order to increase EPO production endogenously. EPO genes have been successfully inserted into mice and monkeys, and were found to increase hematocrits by as much as 80 percent in those animals. However, the endogonous and transgene derived EPO elicited autoimmune responses in some animals in the form of severe anemia.

Insulin-like Growth Factor 1

Insulin-like growth factor 1 is a protein involved in the mediation of the growth hormone. Administration of IGF-1 to mice has resulted in more muscle growth and quicker muscle and nerve regeneration. If athletes were to use this the sustained production of IGF-1 could cause heart disease and cancer.

Others

Modulating the levels of proteins that affect psychology are also potential goals for gene doping; for example pain perception depends on endorphins and enkephalins, response to stress depends on BDNF, and an increase in synthesis of monamines could improve the mood of athletes. Pre-proenkephalin has been administered via gene therapy using a replication-deficient herpes simplex virus, which targets nerves, to mice with results good enough to justify a Phase I clinical trial in people with terminal cancer with uncontrolled pain. Adopting that approach for athletes would be problematic since the pain deadening would likely be permanent.

VEGF has been tested in clinical trials to increase blood flow and has been considered as a potential gene doping agent; however long term follow up of the clinical trial subjects showed poor results. The same is true of fibroblast growth factor. Glucagon-like peptide-1 increases the amount of glucose in the liver and has been administered via gene therapy to the livers of mouse models of diabetes and was shown to increase gluconeogenesis' for athletes this would make more energy available and reduce the buildup of lactic acid.

Detection

The World Anti-Doping Agency (WADA) is the main regulatory organization looking into the issue of the detection of gene doping. Both direct and indirect testing methods are being researched by the organization. Directly detecting the use of gene therapy usually requires the discovery of recombinant proteins or gene insertion vectors, while most indirect methods involve examining the athlete in an attempt to detect bodily changes or structural differences between endogenous and recombinant proteins.

Indirect methods are by nature more subjective, as it becomes very difficult to determine which anomalies are proof of gene doping, and which are simply natural, though unusual, biological properties. For example, Eero Mäntyranta, an Olympic cross country skier, had a mutation which made his body produce abnormally high amounts of red blood cells. It would be very difficult to determine whether or not Mäntyranta's red blood cell levels were due to an innate genetic advantage, or an artificial one.

Research

A 2016 review found that about 120 DNA polymorphisms had been identified in the literature related to some aspect of athletic performance, 77 related to endurance and 43 related to Power. 11 had been replicated in three or more studies and six were identified in genome-wide association studies, but 29 had not been replicated in at least one study.

The 11 replicated markers were:

Endurance

- ACE Alu I/D (rs4646994) (Called ACE I)

- ACTN3 577X

- PPARA rs4253778 G,

- PPARGC1A Gly482;

power/strength markers

- ACE Alu I/D (rs4646994) (called ACE D)

- ACTN3 Arg577

- AMPD1 Gln12

- HIF1A 582Ser

- MTHFR rs1801131 C

- NOS3 rs2070744 T

- PPARG 12Ala

The six GWAS markers were:

- CREM rs1531550 A,

- DMD rs939787 T

- GALNT13 rs10196189 G

- NFIA-AS1 rs1572312 C,

- RBFOX1 rs7191721 G

- TSHR rs7144481 C

Ethics

The mainstream perspective is that gene doping is dangerous and unethical, as is any application of a therapeutic intervention for non-therapeutic or enhancing purposes, and that it compromises the ethical foundation of medicine and the spirit of sport. Others, who support human enhancement on broader grounds, or who see a false dichotomy between "natural" and "artificial" or a denial of the role of technology in improving athletic performance, do not oppose or support gene doping.

Vectors in Gene Therapy

Gene therapy utilizes the delivery of DNA into cells, which can be accomplished by several methods, summarized below. The two major classes of methods are those that use recombinant viruses (sometimes called biological nanoparticles or viral vectors) and those that use naked DNA or DNA complexes (non-viral methods).

Viruses

All viruses bind to their hosts and introduce their genetic material into the host cell as part of their replication cycle. This genetic material contains basic 'instructions' of how to produce more copies of these viruses, hacking the body's normal production machinery to serve the needs of the virus. The host cell will carry out these instructions and produce additional copies of the virus, leading to more and more cells becoming infected. Some types of viruses insert their genome into the host's cytoplasm, but do not actually enter the cell. Others penetrate the cell membrane disguised as protein molecules and enter the cell.

There are two main types of virus infection: lytic and lysogenic. Shortly after inserting its DNA, viruses of the lytic cycle quickly produce more viruses, burst from the cell and infect more cells. Lysogenic viruses integrate their DNA into the DNA of the host cell and may live in the body for many years before responding to a trigger. The virus reproduces as the cell does and does not inflict bodily harm until it is triggered. The trigger releases the DNA from that of the host and employs it to create new viruses.

Retroviruses

The genetic material in retroviruses is in the form of RNA molecules, while the genetic material of their hosts is in the form of DNA. When a retrovirus infects a host cell, it will introduce its RNA together with some enzymes, namely reverse transcriptase and integrase, into the cell. This RNA molecule from the retrovirus must produce a DNA copy from its RNA molecule before it can be integrated into the genetic material of the host cell. The process of producing a DNA copy from an RNA molecule is termed reverse transcription. It is carried out by one of the enzymes carried in the virus, called reverse transcriptase. After this DNA copy is produced and is free in the nucleus of the host cell, it must be incorporated into the genome of the host cell. That is, it must be inserted into the large DNA molecules in the cell (the chromosomes). This process is done by another enzyme carried in the virus called integrase.

Now that the genetic material of the virus has been inserted, it can be said that the host cell has been modified to contain new genes. If this host cell divides later, its descendants will all contain the new genes. Sometimes the genes of the retrovirus do not express their information immediately.

One of the problems of gene therapy using retroviruses is that the integrase enzyme can insert the genetic material of the virus into any arbitrary position in the genome of the host; it randomly inserts the genetic material into a chromosome. If genetic material happens to be inserted in the middle of one of the original genes of the host cell, this gene will be disrupted (insertional mutagenesis). If the gene happens to be one regulating cell division, uncontrolled cell division (i.e., cancer) can occur. This problem has recently begun to be addressed by utilizing zinc finger nucleases or by including certain sequences such as the beta-globin locus control region to direct the site of integration to specific chromosomal sites.

Gene therapy trials using retroviral vectors to treat X-linked severe combined immunodeficiency (X-SCID) represent the most successful application of gene therapy to date. More than twenty patients have been treated in France and Britain, with a high rate of immune system reconstitution observed. Similar trials were restricted or halted in the USA when leukemia was reported in patients treated in the French X-SCID gene therapy trial. To date, four children in the French trial and one in the British trial have developed leukemia as a result of insertional mutagenesis by the retroviral vector. All but one of these children responded well to conventional anti-leukemia treatment. Gene therapy trials to treat SCID due to deficiency of the Adenosine Deaminase (ADA) enzyme (one form of SCID) continue with relative success in the USA, Britain, Ireland, Italy and Japan.

Adenoviruses

Adenoviruses are viruses that carry their genetic material in the form of double-stranded DNA. They cause respiratory, intestinal, and eye infections in humans (especially the common cold). When these viruses infect a host cell, they introduce their DNA molecule into the host. The genetic material of the adenoviruses is not incorporated (transient) into the host cell's genetic material. The DNA molecule is left free in the nucleus of the host cell, and the instructions in this extra DNA molecule are transcribed just like any other gene. The only difference is that these extra genes are not replicated when the cell is about to undergo cell division so the descendants of that cell will not have the extra gene. As a result, treatment with the adenovirus will require readministration in a growing cell population although the absence of integration into the host cell's genome should prevent the type of cancer seen in the SCID trials. This vector system has been promoted for treating cancer and indeed the first gene therapy product to be licensed to treat cancer, Gendicine, is an adenovirus. Gendicine, an adenoviral p53-based gene therapy was approved by the Chinese food and drug regulators in 2003 for treatment of head and neck cancer. Advexin, a similar gene therapy approach from Introgen, was turned down by the US Food and Drug Administration (FDA) in 2008.

Concerns about the safety of adenovirus vectors were raised after the 1999 death of Jesse Gelsinger while participating in a gene therapy trial. Since then, work using adenovirus vectors has focused on genetically crippled versions of the virus.

Envelope Protein Pseudotyping of Viral Vectors

The viral vectors described above have natural host cell populations that they infect most efficiently. Retroviruses have limited natural host cell ranges, and although adenovirus and adeno-associated virus are able to infect a relatively broader range of cells efficiently, some cell types are refractory to infection by these viruses as well. Attachment to and entry into a susceptible cell is mediated by the protein envelope on the surface of a virus. Retroviruses and adeno-associated viruses have a single protein coating their membrane, while adenoviruses are coated with both an envelope protein and fibers that extend away from the surface of the virus. The envelope proteins on each of these viruses bind to cell-surface molecules such as heparin sulfate, which localizes them upon the surface of the potential host, as well as with the specific protein receptor that either induces entry-promoting structural changes in the viral protein, or localizes the virus in endosomes wherein acidification of the lumen induces this refolding of the viral coat. In either case, entry into potential host cells requires a favorable interaction between a protein on the surface of the virus and a protein on the surface of the cell. For the purposes of gene therapy, one might either want to limit or expand the range of cells susceptible to transduction by a gene therapy vector. To this end, many vectors have been developed in which the endogenous viral envelope proteins have been replaced by either envelope proteins from other viruses, or by chimeric proteins. Such chimera would consist of those parts of the viral protein necessary for incorporation into the virion as well as sequences meant to interact with specific host cell proteins. Viruses in which the envelope proteins have been replaced as described are referred to as pseudotyped viruses. For example, the most popular retroviral vector for use in gene therapy trials has been the lentivirus Simian immunodeficiency virus coated with the envelope proteins, G-protein, from Vesicular stomatitis virus. This vector is referred to as VSV G-pseudotyped lentivirus, and infects an almost universal set of cells. This tropism is characteristic of the VSV G-protein with which this vector is coated. Many attempts have been made to limit the tropism of viral vectors to one or a few host cell populations. This advance would allow for the systemic administration of a relatively small amount of vector. The potential for off-target cell modification would be limited, and many concerns from the medical community would be alleviated. Most attempts to limit tropism have used chimeric envelope proteins bearing antibody fragments. These vectors show great promise for the development of "magic bullet" gene therapies.

Replication-competent Vectors

A replication-competent vector called ONYX-015 is used in replicating tumor cells. It was found that in the absence of the E1B-55Kd viral protein, adenovirus caused very rapid apoptosis of infected, p53(+) cells, and this results in dramatically reduced virus progeny and no subsequent spread.

Apoptosis was mainly the result of the ability of EIA to inactivate p300. In p53(-) cells, deletion of E1B 55kd has no consequence in terms of apoptosis, and viral replication is similar to that of wild-type virus, resulting in massive killing of cells.

A replication-defective vector deletes some essential genes. These deleted genes are still necessary in the body so they are replaced with either a helper virus or a DNA molecule.

Cis and Trans-acting Elements

Replication-defective vectors always contain a "transfer construct". The transfer construct carries the gene to be transduced or "transgene". The transfer construct also carries the sequences which are necessary for the general functioning of the viral genome: packaging sequence, repeats for replication and, when needed, priming of reverse transcription. These are denominated cis-acting elements, because they need to be on the same piece of DNA as the viral genome and the gene of interest. Trans-acting elements are viral elements, which can be encoded on a different DNA molecule. For example, the viral structural proteins can be expressed from a different genetic element than the viral genome.

Herpes Simplex Virus

The Herpes simplex virus is a human neurotropic virus. This is mostly examined for gene transfer in the nervous system. The wild type HSV-1 virus is able to infect neurons and evade the host immune response, but may still become reactivated and produce a lytic cycle of viral replication. Therefore, it is typical to use mutant strains of HSV-1 that are deficient in their ability to replicate. Though the latent virus is not transcriptionally apparent, it does possess neuron specific promoters that can continue to function normally. Antibodies to HSV-1 are common in humans, however complications due to herpes infection are somewhat rare. Caution for rare cases of encephalitis must be taken and this provides some rationale to using HSV-2 as a viral vector as it generally has tropism for neuronal cells innervating the urogenital area of the body and could then spare the host of severe pathology in the brain.

Non-viral Methods

Non-viral methods present certain advantages over viral methods, with simple large scale production and low host immunogenicity being just two. Previously, low levels of transfection and expression of the gene held non-viral methods at a disadvantage; however, recent advances in vector technology have yielded molecules and techniques with transfection efficiencies similar to those of viruses.

Injection of Naked DNA

This is the simplest method of non-viral transfection. Clinical trials carried out of intramuscular injection of a naked DNA plasmid have occurred with some success; however, the expression has been very low in comparison to other methods of transfection. In addition to trials with plasmids, there have been trials with naked PCR product, which have had similar or greater success. Cellular uptake of naked DNA is generally inefficient. Research efforts focusing on improving the efficiency of naked DNA uptake have yielded several novel methods, such as electroporation, sonoporation, and the use of a "gene gun", which shoots DNA coated gold particles into the cell using high pressure gas.

Physical Methods to Enhance Delivery

Electroporation

Electroporation is a method that uses short pulses of high voltage to carry DNA across the cell membrane. This shock is thought to cause temporary formation of pores in the cell membrane, allowing DNA molecules to pass through. Electroporation is generally efficient and works across a broad range of cell types. However, a high rate of cell death following electroporation has limited its use, including clinical applications.

More recently a newer method of electroporation, termed electron-avalanche transfection, has been used in gene therapy experiments. By using a high-voltage plasma discharge, DNA was efficiently delivered following very short (microsecond) pulses. Compared to electroporation, the technique resulted in greatly increased efficiency and less cellular damage.

Gene Gun

The use of particle bombardment, or the gene gun, is another physical method of DNA transfection. In this technique, DNA is coated onto gold particles and loaded into a device which generates a force to achieve penetration of the DNA into the cells, leaving the gold behind on a "stopping" disk.

Sonoporation

Sonoporation uses ultrasonic frequencies to deliver DNA into cells. The process of acoustic cavitation is thought to disrupt the cell membrane and allow DNA to move into cells.

Magnetofection

In a method termed magnetofection, DNA is complexed to magnetic particles, and a magnet is placed underneath the tissue culture dish to bring DNA complexes into contact with a cell monolayer.

Hydrodynamic Delivery

Hydrodynamic delivery involves rapid injection of a high volume of a solution into vasculature (such as into the inferior vena cava, bile duct, or tail vein). The solution contains molecules that are to be inserted into cells, such as DNA plasmids or siRNA, and transfer of these molecules into cells is assisted by the elevated hydrostatic pressure caused by the high volume of injected solution.

Chemical Methods to Enhance Delivery

Oligonucleotides

The use of synthetic oligonucleotides in gene therapy is to deactivate the genes involved in the disease process. There are several methods by which this is achieved. One strategy uses antisense specific to the target gene to disrupt the transcription of the faulty gene. Another uses small molecules of RNA called siRNA to signal the cell to cleave specific unique sequences in the mRNA transcript of the faulty gene, disrupting translation of the faulty mRNA, and therefore expression of the gene. A further strategy uses double stranded oligodeoxynucleotides as a decoy for the transcription factors that are required to activate the transcription of the target gene. The transcription factors bind to the decoys instead of the promoter of the faulty gene, which reduces the transcription of the target gene, lowering expression. Additionally, single stranded DNA oligonucleotides have been used to direct a single base change within a mutant gene. The oligonucleotide is designed to anneal with complementarity to the target gene with the exception of a central base, the target base, which serves as the template base for repair. This technique is referred to as oligonucleotide mediated gene repair, targeted gene repair, or targeted nucleotide alteration.

Lipoplexes

To improve the delivery of the new DNA into the cell, the DNA must be protected from damage and (positively charged). Initially, anionic and neutral lipids were used for the construction of lipoplexes for synthetic vectors. However, in spite of the facts that there is little toxicity associated with them, that they are compatible with body fluids and that there was a possibility of adapting them to be tissue specific; they are complicated and time consuming to produce so attention was turned to the cationic versions.

Cationic lipids, due to their positive charge, were first used to condense negatively charged DNA molecules so as to facilitate the encapsulation of DNA into liposomes. Later it was found that the use of cationic lipids significantly enhanced the stability of lipoplexes. Also as a result of their charge, cationic liposomes interact with the cell membrane, endocytosis was widely believed as the major route by which cells uptake lipoplexes. Endosomes are formed as the results of endocytosis, however, if genes can not be released into cytoplasm by breaking the membrane of endosome, they will be sent to lysosomes where all DNA will be destroyed before they could achieve their functions. It was also found that although cationic lipids themselves could condense and encapsulate DNA into liposomes, the transfection efficiency is very low due to the lack of ability in terms of "endosomal escaping". However, when helper lipids (usually electroneutral lipids, such as DOPE) were added to form lipoplexes, much higher transfection efficiency was observed. Later on, it was figured out that certain lipids have the ability to destabilize endosomal membranes so as to facilitate the escape of DNA from endosome, therefore those lipids are called fusogenic lipids. Although cationic liposomes have been widely used as an alternative for gene delivery vectors, a dose dependent toxicity of cationic lipids were also observed which could limit their therapeutic usages.

The most common use of lipoplexes has been in gene transfer into cancer cells, where the supplied genes have activated tumor suppressor control genes in the cell and decrease the activity of oncogenes. Recent studies have shown lipoplexes to be useful in transfecting respiratory epithelial cells.

Polymersomes

Polymersomes are synthetic versions of liposomes (vesicles with a lipid bilayer), made of amphiphilic block copolymers. They can encapsulate either hydrophilic or hydrophobic contents and can be used to deliver cargo such as DNA, proteins, or drugs to cells. Advantages of polymersomes over liposomes include greater stability, mechanical strength, blood circulation time, and storage capacity.

Polyplexes

Complexes of polymers with DNA are called polyplexes. Most polyplexes consist of cationic polymers and their fabrication is based on self-assembly by ionic interactions. One important difference between the methods of action of polyplexes and lipoplexes is that polyplexes cannot directly release their DNA load into the cytoplasm. As a result, co-transfection with endosome-lytic agents such as inactivated adenovirus was required to facilitate nanoparticle escape from the endocytic vesicle made during particle uptake. However, a better understanding of the mechanisms by which DNA can escape from endolysosomal pathway, i.e. proton sponge effect, has triggered new polymer synthesis strategies such as incorporation of protonable residues in polymer backbone and has revitalized research on polycation-based systems.

Due to their low toxicity, high loading capacity, and ease of fabrication, polycationic nanocarriers demonstrate great promise compared to their rivals such as viral vectors which show high immunogenicity and potential carcinogenicity, and lipid-based vectors which cause dose dependence toxicity. Polyethyleneimine and chitosan are among the polymeric carriers that have been extensively studies for development of gene delivery therapeutics. Other polycationic carriers such as poly(beta-amino esters) and polyphosphoramidate are being added to the library of potential gene carriers. In addition to the variety of polymers and copolymers, the ease of controlling the size, shape, surface chemistry of these polymeric nano-carriers gives them an edge in targeting capability and taking advantage of enhanced permeability and retention effect.

Dendrimers

A dendrimer is a highly branched macromolecule with a spherical shape. The surface of the particle may be functionalized in many ways and many of the properties of the resulting construct are determined by its surface.

In particular it is possible to construct a cationic dendrimer, i.e. one with a positive surface charge. When in the presence of genetic material such as DNA or RNA, charge complimentarity leads to a temporary association of the nucleic acid with the cationic dendrimer. On reaching its destination the dendrimer-nucleic acid complex is then taken into the cell via endocytosis.

In recent years the benchmark for transfection agents has been cationic lipids. Limitations of these competing reagents have been reported to include: the lack of ability to transfect some cell types, the lack of robust active targeting capabilities, incompatibility with animal models, and toxicity. Dendrimers offer robust covalent construction and extreme control over molecule structure, and therefore size. Together these give compelling advantages compared to existing approaches.

Producing dendrimers has historically been a slow and expensive process consisting of numerous slow reactions, an obstacle that severely curtailed their commercial development. The Michigan-based company Dendritic Nanotechnologies discovered a method to produce dendrimers using kinetically driven chemistry, a process that not only reduced cost by a magnitude of three, but also cut reaction time from over a month to several days. These new "Priostar" dendrimers can be specifically constructed to carry a DNA or RNA payload that transfects cells at a high efficiency with little or no toxicity.

Inorganic Nanoparticles

Inorganic nanoparticles, such as gold, silica, iron oxide (ex. magnetofection) and calcium phosphates have been shown to be capable of gene delivery. Some of the benefits of inorganic vectors is in their storage stability, low manufacturing cost and often time, low immunogenicity, and resistance to microbial attack. Nanosized materials less than 100 nm have been shown to efficiently trap the DNA or RNA and allows its escape from the endosome without degradation. Inorganics have also been shown to exhibit improved in vitro transfection for attached cell lines due to their increased density and preferential location on the base of the culture dish. Quantum dots have also been used successfully and permits the coupling of gene therapy with a stable fluorescence marker.

Cell-penetrating Peptides

Cell-penetrating peptides (CPPs), also known as peptide transduction domains (PTDs), are short peptides (< 40 amino acids) that efficiently pass through cell membranes while being covalently or non-covalently bound to various molecules, thus facilitating these molecules' entry into cells. Cell entry occurs primarily by endocytosis but other entry mechanisms also exist. Examples of cargo molecules of CPPs include nucleic acids, liposomes, and drugs of low molecular weight.

CPP cargo can be directed into specific cell organelles by incorporating localization sequences into CPP sequences. For example, nuclear localization sequences are commonly used to guide CPP cargo into the nucleus. For guidance into mitochondria, a mitochondrial targeting sequence can be used; this method is used in protofection (a technique that allows for foreign mitochondrial DNA to be inserted into cells' mitochondria).

Hybrid Methods

Due to every method of gene transfer having shortcomings, there have been some hybrid methods developed that combine two or more techniques. Virosomes are one example; they combine liposomes with an inactivated HIV or influenza virus. This has been shown to have more efficient gene transfer in respiratory epithelial cells than either viral or liposomal methods alone. Other methods involve mixing other viral vectors with cationic lipids or hybridising viruses.

References

- Soliday, F. K.; Conley, Y. P.; Henker, R. (2010). "Pseudocholinesterase deficiency: A comprehensive review of genetic, acquired, and drug influences". AANA Journal. 78 (4): 313–320. PMID 20879632

- Permezel, Michael; Walker, Susan; Kyprianou, Kypros (2015). Beischer & MacKay's Obstetrics, Gynaecology and the Newborn. Elsevier Health Sciences. p. 74. ISBN 9780729584050. Retrieved 24 January 2017

- "Genetic Testing for Hereditary Cancer Syndromes". National Cancer Institute. National Institute of Health. Retrieved 18 November 2016

- Holtzman NA; Murphy PD; Watson MS; Barr PA (October 1997). "Predictive genetic testing: from basic research to clinical practice". Science. 278 (5338): 602–5. PMID 9381169. doi:10.1126/science.278.5338.602

- Borry, P., Cornel, M., HOWARD, H. (2011). Where are you going, where have you been. A recent history of the direct-to-consumer genetic testing market. Journal of Community Genetics

- Gibson, Greg; Muse, Spencer V (2009). A Primer of Genome Science (Third ed.). 23 Plumtree Rd, Sunderland, MA 01375: Sinauer Associates. pp. 304–305. ISBN 978-0-87893-236-8

- "Family Tree DNA reaches a historic milestone: over 500,000 DNA tests" (PDF) (Press release). Family Tree DNA. Retrieved November 29, 2012

- Porteus, M (2016). "Genome Editing: A New Approach to Human Therapeutics". Annual review of pharmacology and toxicology. 56: 163–90. PMID 26566154. doi:10.1146/annurev-pharmtox-010814-124454

- "Supreme Court Ruling Today Allows DNATraits to Offer Low Cost BRCA Breast and Ovarian Cancer Gene Testing in U.S.". Wall Street Journal. Wall Street Journal. Retrieved June 19, 2013

- Lodish, H; Berk, A; Zipursky, SL; et. al (2000). Molecular Cell Biology (Fourth ed.). New York: W. H. Freeman and Company. pp. Section 6.3, Viruses: Structure, Function, and Uses. ISBN 9780716737063

- Bettinger, Blaine (May 28, 2008). "Interview Series I – Bennett Greenspan of Family Tree DNA". The Genetic Genealogist. Retrieved 24 June 2013

- Galanello, R.; Origa, R. (2010). "Beta-thalassemia". Orphanet Journal of Rare Diseases. 5: 11. PMC 2893117. PMID 20492708. doi:10.1186/1750-1172-5-11

- Strachnan, T.; Read, A. P. (2004). Human Molecular Genetics (3rd ed.). Garland Publishing. p. 616. ISBN 0815341849

- Fischer, A.; Hacein-Bey-Abina, S.; Cavazzana-Calvo, M. (2010). "20 years of gene therapy for SCID". Nature Immunology. 11 (6): 457–460. PMID 20485269. doi:10.1038/ni0610-457

- Allhoff, F. (2005). "Germ-Line Genetic Enhancement and Rawlsian Primary Goods". Kennedy Institute of Ethics Journal. 15 (1): 39–56. PMID 15881795. doi:10.1353/ken.2005.0007

- Wilson, Jennifer Fisher (18 March 2002). "Murine Gene Therapy Corrects Symptoms of Sickle Cell Disease". The Scientist – Magazine of the Life Sciences. Retrieved 17 August 2010

- Ahmetov, I. I.; Fedotovskaya, O. N. (2015). "Current Progress in Sports Genomics". Advances in Clinical Chemistry. Advances in Clinical Chemistry. 70: 247–314. ISBN 9780128033166. PMID 26231489. doi:10.1016/bs.acc.2015.03.003. review

- Birzniece, V (2015). "Doping in sport: Effects, harm and misconceptions". Internal Medicine Journal. 45 (3): 239–48. PMID 25369881. doi:10.1111/imj.12629

- Nayerossadat, Nouri; Maedeh, Talebi; Abas, Ali (6 July 2012). "Viral and nonviral delivery systems for gene delivery". Advanced Biomedical Research. 1: 27. PMC 3507026. doi:10.4103/2277-9175.98152

- Reissman, S. (2014). "Cell penetration: scope and limitations by the application of cell-penetrating peptides.". Journal of Peptide Science. 20 (10): 760–784. doi:10.1002/psc.2672

- Dvorsky, George (6 March 2014) Scientists Create Genetically Modified Cells That Protect Against HIV io9, Biotechnology. Retrieved 6 March 2014

- Cyranoski, David (28 July 2016). "Chinese scientists to pioneer first human CRISPR trial". Nature. 535 (7613): 476–477. PMID 27466105. doi:10.1038/nature.2016.20302

6

Understanding Transformation Genetics

Transformation is the alteration of cells resulting from the incorporation of genetic material from its natural surroundings. For the process of transformation to take place, the bacteria must be in a state of competence. The subject deals with themes like natural competence, agrobacterium, viral transformation and transformation efficiency. This chapter will provide an integrated understanding of transformation genetics.

Transformation (Genetics)

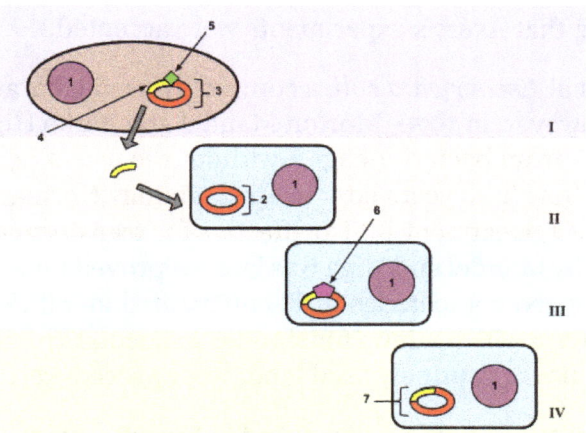

In this image, a gene from bacterial cell 1 is moved from bacterial cell 1 to bacterial cell 2.
This process of bacterial cell 2 taking up new genetic material is called transformation.

In molecular biology, transformation is the genetic alteration of a cell resulting from the direct uptake and incorporation of exogenous genetic material from its surroundings through the cell membrane(s). For transformation to take place, the recipient bacteria must be in a state of competence, which might occur in nature as a time-limited response to environmental conditions such as starvation and cell density, and may also be induced in a laboratory.

Transformation is one of three processes for horizontal gene transfer, in which exogenous genetic material passes from bacterium to another, the other two being conjugation (transfer of genetic material between two bacterial cells in direct contact) and transduction (injection of foreign DNA by a bacteriophage virus into the host bacterium). In transformation, the genetic material passes through the intervening medium, and uptake is completely dependent on the recipient bacterium.

As of 2014 about 80 species of bacteria were known to be capable of transformation, about evenly divided between Gram-positive and Gram-negative bacteria; the number might be an overestimate since several of the reports are supported by single papers.

"Transformation" may also be used to describe the insertion of new genetic material into non-bacterial cells, including animal and plant cells; however, because "transformation" has a special meaning in relation to animal cells, indicating progression to a cancerous state, the process is usually called "transfection".

History

Transformation in bacteria was first demonstrated in 1928 by British bacteriologist Frederick Griffith. Griffith discovered that a strain of *Streptococcus pneumoniae* could be made virulent after being exposed to heat-killed virulent strains. Griffith hypothesized that some "transforming principle" from the heat-killed strain was responsible for making the harmless strain virulent. In 1944 this "transforming principle" was identified as being genetic by Oswald Avery, Colin MacLeod, and Maclyn McCarty. They isolated DNA from a virulent strain of *S. pneumoniae* and using just this DNA were able to make a harmless strain virulent. They called this uptake and incorporation of DNA by bacteria "transformation" . The results of Avery et al.'s experiments were at first skeptically received by the scientific community and it was not until the development of genetic markers and the discovery of other methods of genetic transfer (conjugation in 1947 and transduction in 1953) by Joshua Lederberg that Avery's experiments were accepted.

It was originally thought that *Escherichia coli*, a commonly used laboratory organism, was refractory to transformation. However, in 1970, Morton Mandel and Akiko Higa showed that *E. coli* may be induced to take up DNA from bacteriophage λ without the use of helper phage after treatment with calcium chloride solution. Two years later in 1972, Stanley Norman Cohen, Annie Chang and Leslie Hsu showed that $CaCl_2$ treatment is also effective for transformation of plasmid DNA. The method of transformation by Mandel and Higa was later improved upon by Douglas Hanahan. The discovery of artificially induced competence in *E. coli* created an efficient and convenient procedure for transforming bacteria which allows for simpler molecular cloning methods in biotechnology and research, and it is now a routinely used laboratory procedure.

Transformation using electroporation was developed in the late 1980s, increasing the efficiency of in-vitro transformation and increasing the number of bacterial strains that could be transformed. Transformation of animal and plant cells was also investigated with the first transgenic mouse being created by injecting a gene for a rat growth hormone into a mouse embryo in 1982. In 1907 a bacterium that caused plant tumors, *Agrobacterium tumefaciens*, was discovered and in the early 1970s the tumor-inducing agent was found to be a DNA plasmid called the Ti plasmid. By removing the genes in the plasmid that caused the tumor and adding in novel genes, researchers were able to infect plants with *A. tumefaciens* and let the bacteria insert their chosen DNA into the genomes of the plants. Not all plant cells are susceptible to infection by *A. tumefaciens*, so other methods were developed, including electroporation and micro-injection. Particle bombardment was made possible with the invention of the Biolistic Particle Delivery System (gene gun) by John Sanford in the 1980s.

Definitions

Transformation is one of three forms of horizontal gene transfer that occur in nature among bacteria, in which DNA encoding for a trait passes from one bacterium to another and is integrated into the recipient genome by homologous recombination; the other two are transduction, carried out by means of a bacteriophage, and conjugation, in which a gene is passed through direct contact between bacteria. In transformation, the genetic material passes through the intervening medium, and uptake is completely dependent on the recipient bacterium.

Competence refers to a temporary state of being able to take up exogenous DNA from the environment; it may be induced in a laboratory.

It appears to be an ancient process inherited from a common prokaryotic ancestor that is a beneficial adaptation for promoting recombinational repair of DNA damage, especially damage acquired under stressful conditions. Natural genetic transformation appears to be an adaptation for repair of DNA damage that also generates genetic diversity.

Transformation has been studied in medically important Gram-negative bacteria species such as *Helicobacter pylori*, *Legionella pneumophila*, *Neisseria meningitidis*, *Neisseria gonorrhoeae*, *Haemophilus influenzae* and *Vibrio cholerae*. It has also been studied in Gram-negative species found in soil such as *Pseudomonas stutzeri*, *Acinetobacter baylyi*, and Gram-negative plant pathogens such as *Ralstonia solanacearum* and *Xylella fastidiosa*. Transformation among Gram-positive bacteria has been studied in medically important species such as *Streptococcus pneumoniae*, *Streptococcus mutans*, *Staphylococcus aureus* and *Streptococcus sanguinis* and in Gram-positive soil bacterium *Bacillus subtilis*. It has also been reported in at least 30 species of *Proteobacteria* distributed in the classes alpha, beta, gamma and epsilon. The best studied *Proteobacteria* with respect to transformation are the medically important human pathogens *Neisseria gonorrhoeae* (class beta), *Haemophilus influenzae* (class gamma) and *Helicobacter pylori* (class epsilon)

"Transformation" may also be used to describe the insertion of new genetic material into non-bacterial cells, including animal and plant cells; however, because "transformation" has a special meaning in relation to animal cells, indicating progression to a cancerous state, the process is usually called "transfection".

Natural Competence and Transformation

As of 2014 about 80 species of bacteria were known to be capable of transformation, about evenly divided between Gram-positive and Gram-negative bacteria; the number might be an overestimate since several of the reports are supported by single papers.

Naturally competent bacteria carry sets of genes that provide the protein machinery to bring DNA across the cell membrane(s). The transport of the exogenous DNA into the cells may require proteins that are involved in the assembly of type IV pili and type II secretion system, as well as DNA translocase complex at the cytoplasmic membrane.

Due to the differences in structure of the cell envelope between Gram-positive and Gram-negative bacteria, there are some differences in the mechanisms of DNA uptake in these cells, however most of them share common features that involve related proteins. The DNA first binds to the surface of the competent cells on a DNA receptor, and passes through the cytoplasmic membrane via DNA translocase. Only single-stranded DNA may pass through, the other strand being degraded by nucleases in the process. The translocated single-stranded DNA may then be integrated into the bacterial chromosomes by a RecA-dependent process. In Gram-negative cells, due to the presence of an extra membrane, the DNA requires the presence of a channel formed by secretins on the out-er membrane. Pilin may be required for competence, but its role is uncertain. The uptake of DNA is generally non-sequence specific, although in some species the presence of specific DNA uptake sequences may facilitate efficient DNA uptake.

Natural Transformation

Natural transformation is a bacterial adaptation for DNA transfer that depends on the expression of numerous bacterial genes whose products appear to be responsible for this process. In general, transformation is a complex, energy-requiring developmental process. In order for a bacterium to bind, take up and recombine exogenous DNA into its chromosome, it must become competent, that is, enter a special physiological state. Competence development in *Bacillus subtilis* requires expression of about 40 genes. The DNA integrated into the host chromosome is usually (but with rare exceptions) derived from another bacterium of the same species, and is thus homologous to the resident chromosome.

In *B. subtilis* the length of the transferred DNA is greater than 1271 kb (more than 1 million bases). The length transferred is likely double stranded DNA and is often more than a third of the total chromosome length of 4215 kb. It appears that about 7-9% of the recipient cells take up an entire chromosome.

The capacity for natural transformation appears to occur in a number of prokaryotes, and thus far 67 prokaryotic species (in seven different phyla) are known to undergo this process.

Competence for transformation is typically induced by high cell density and/or nutritional limitation, conditions associated with the stationary phase of bacterial growth. Transformation in *Haemophilus influenzae* occurs most efficiently at the end of exponential growth as bacterial growth approaches stationary phase. Transformation in *Streptococcus mutans*, as well as in many other streptococci, occurs at high cell density and is associated with biofilm formation. Competence in *B. subtilis* is induced toward the end of logarithmic growth, especially under conditions of amino acid limitation.

By releasing intact host and plasmid DNA, certain bacteriophages are thought to contribute to transformation.

Transformation, as an Adaptation for DNA Repair

Competence is specifically induced by DNA damaging conditions. For instance, transformation is induced in *Streptococcus pneumoniae* by the DNA damaging agents mitomycin C (a DNA cross-linking agent) and fluoroquinolone (a topoisomerase inhibitor that causes double-strand breaks). In *B. subtilis*, transformation is increased by UV light, a DNA damaging agent. In *Helicobacter pylori*, ciprofloxacin, which interacts with DNA gyrase and introduces double-strand breaks, induces expression of competence genes, thus enhancing the frequency of transformation Using *Legionella pneumophila*, Charpentier et al. tested 64 toxic molecules to determine which of these induce competence. Of these only six, all DNA damaging agents, caused strong induction. These DNA damaging agents were mitomycin C (which causes DNA inter-strand crosslinks), norfloxacin, ofloxacin and nalidixic acid (inhibitors of DNA gyrase that cause double-strand breaks), bicyclomycin (causes single- and double-strand breaks), and hydroxyurea (induces DNA base oxidation). UV light also induced competence in *L. pneumophila*. Charpentier et al. suggested that competence for transformation probably evolved as a DNA damage response.

Logarithmically growing bacteria differ from stationary phase bacteria with respect to the number of genome copies present in the cell, and this has implications for the capability to carry out an important DNA repair process. During logarithmic growth, two or more copies of any particular region of the chromosome may be present in a bacterial cell, as cell division is not precisely matched with chromosome replication. The process of homologous recombinational repair (HRR) is a key DNA repair process that is especially effective for repairing double-strand damages, such as double-strand breaks. This process depends on a second homologous chromosome in addition to the damaged chromosome. During logarithmic growth, a DNA damage in one chromosome may be repaired by HRR using sequence information from the other homologous chromosome. Once cells approach stationary phase, however, they typically have just one copy of the chromosome, and HRR requires input of homologous template from outside the cell by transformation.

To test whether the adaptive function of transformation is repair of DNA damages, a series of experiments were carried out using *B. subtilis* irradiated by UV light as the damaging agent (reviewed by Michod et al. and Bernstein et al.) The results of these experiments indicated that transforming DNA acts to repair potentially lethal DNA damages introduced by UV light in the recipient DNA. The particular process responsible for repair was likely HRR. Transformation in bacteria can be viewed as a primitive sexual process, since it involves interaction of homologous DNA from two individuals to form recombinant DNA that is passed on to succeeding generations. Bacterial transformation in prokaryotes may have been the ancestral process that gave rise to meiotic sexual reproduction in eukaryotes.

Methods and Mechanisms of Transformation in Laboratory

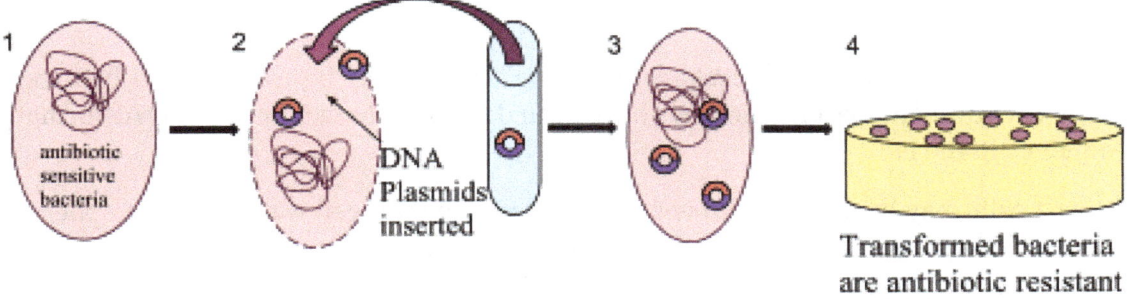

Schematic of bacterial transformation – for which artificial competence must first be induced.

Bacterial

Artificial competence can be induced in laboratory procedures that involve making the cell passively permeable to DNA by exposing it to conditions that do not normally occur in nature. Typically the cells are incubated in a solution containing divalent cations (often calcium chloride) under cold conditions, before being exposed to a heat pulse (heat shock). Calcium chloride partially disrupts the cell membrane, which allows the recombinant DNA enter the host cell. Cells that are able to take up the DNA are called competent cells.

It has been found that growth of Gram-negative bacteria in 20 mM Mg reduces the number of protein-to-lipopolysaccharide bonds by increasing the ratio of ionic to covalent bonds, which increases membrane fluidity, facilitating transformation. The role of lipopolysaccharides here are verified from the observation that shorter O-side chains are more effectively transformed – perhaps because of improved DNA accessibility.

The surface of bacteria such as *E. coli* is negatively charged due to phospholipids and lipopolysaccharides on its cell surface, and the DNA is also negatively charged. One function of the divalent cation therefore would be to shield the charges by coordinating the phosphate groups and other negative charges, thereby allowing a DNA molecule to adhere to the cell surface.

DNA entry into *E. coli* cells is through channels known as zones of adhesion or Bayer's junction, with a typical cell carrying as many as 400 such zones. Their role was established when cobalamine (which also uses these channels) was found to competitively inhibit DNA uptake. Another type of channel implicated in DNA uptake consists of poly (HB):poly P:Ca. In this poly (HB) is envisioned to wrap around DNA (itself a polyphosphate), and is carried in a shield formed by Ca ions.

It is suggested that exposing the cells to divalent cations in cold condition may also change or weaken the cell surface structure, making it more permeable to DNA. The heat-pulse is thought to create a thermal imbalance across the cell membrane, which forces the DNA to enter the cells through either cell pores or the damaged cell wall.

Electroporation is another method of promoting competence. In this method the cells are briefly shocked with an electric field of 10-20 kV/cm, which is thought to create holes in the cell membrane through which the plasmid DNA may enter. After the electric shock, the holes are rapidly closed by the cell's membrane-repair mechanisms.

Yeast

Most species of yeast, including *Saccharomyces cerevisiae*, may be transformed by exogenous DNA in the environment. Several methods have been developed to facilitate this transformation at high frequency in the lab.

- Yeast cells may be treated with enzymes to degrade their cell walls, yielding spheroplasts. These cells are very fragile but take up foreign DNA at a high rate.

- Exposing intact yeast cells to alkali cations such as those of cesium or lithium allows the cells to take up plasmid DNA. Later protocols adapted this transformation method, using lithium acetate, polyethylene glycol, and single-stranded DNA. In these protocols, the single-stranded DNA preferentially binds to the yeast cell wall, preventing plasmid DNA from doing so and leaving it available for transformation.

- Electroporation: Formation of transient holes in the cell membranes using electric shock; this allows DNA to enter as described above for bacteria.

- Enzymatic digestion or agitation with glass beads may also be used to transform yeast cells.

Efficiency – Different yeast genera and species take up foreign DNA with different efficiencies.

Also, most transformation protocols have been developed for baker's yeast, *S. cerevisiae*, and thus may not be optimal for other species. Even within one species, different strains have different transformation efficiencies, sometimes different by three orders of magnitude. For instance, when S. cerevisiae strains were transformed with 10 ug of plasmid YEp13, the strain DKD-5D-H yielded between 550 and 3115 colonies while strain OS1 yielded fewer than five colonies.

Plants

A number of methods are available to transfer DNA into plant cells. Some vector-mediated methods are:

- *Agrobacterium*-mediated transformation is the easiest and most simple plant transformation. Plant tissue (often leaves) are cut into small pieces, e.g. 10x10mm, and soaked for 10 minutes in a fluid containing suspended *Agrobacterium*. The bacteria will attach to many of the plant cells exposed by the cut. The plant cells secrete wound-related phenolic compounds which in turn act to upregulate the virulence operon of the Agrobacterium. The virulence operon includes many genes that encode for proteins that are part of a Type IV secretion system that exports from the bacterium proteins and DNA (delineated by specific recognition motifs called border sequences and excised as a single strand from the virulence plasmid) into the plant cell through a structure called a pilus. The transferred DNA (called T-DNA) is piloted to the plant cell nucleus by nuclear localization signals present in the Agrobacterium protein VirD2, which is covalently attached to the end of the T-DNA at the Right border (RB). Exactly how the T-DNA is integrated into the host plant genomic DNA is an active area of plant biology research. Assuming that a selection marker (such as an antibiotic resistance gene) was included in the T-DNA, the transformed plant tissue can be cultured on selective media to produce shoots. The shoots are then transferred to a different medium to promote root formation. Once roots begin to grow from the transgenic shoot, the plants can be transferred to soil to complete a normal life cycle (make seeds). The seeds from this first plant (called the T1, for first transgenic generation) can be planted on a selective (containing an antibiotic), or if an herbicide resistance gene was used, could alternatively be planted in soil, then later treated with herbicide to kill wildtype segregants. Some plants species, such as *Arabidopsis thaliana* can be transformed by dipping the flowers or whole plant, into a suspension of *Agrobacterium tumefaciens*, typically strain C58 (C=Cherry, 58=1958, the year in which this particular strain of *A. tumefaciens* was isolated from a cherry tree in an orchard at Cornell University in Ithaca, New York). Though many plants remain recalcitrant to transformation by this method, research is ongoing that continues to add to the list the species that have been successfully modified in this manner.

- Viral transformation (transduction): Package the desired genetic material into a suitable plant virus and allow this modified virus to infect the plant. If the genetic material is DNA, it can recombine with the chromosomes to produce transformant cells. However, genomes of most plant viruses consist of single stranded RNA which replicates in the cytoplasm of infected cell. For such genomes this method is a form of transfection and not a real transformation, since the inserted genes never reach the nucleus of the cell and do not integrate into the host genome. The progeny of the infected plants is virus-free and also free of the inserted gene.

Some vector-less methods include:

- Gene gun: Also referred to as particle bombardment, microprojectile bombardment, or biolistics. Particles of gold or tungsten are coated with DNA and then shot into young plant cells or plant embryos. Some genetic material will stay in the cells and transform them. This method also allows transformation of plant plastids. The transformation efficiency is lower than in *Agrobacterium*-mediated transformation, but most plants can be transformed with this method.

- Electroporation: Formation of transient holes in cell membranes using electric pulses of high field strength; this allows DNA to enter as described above for bacteria.

Animals

Introduction of DNA into animal cells is usually called transfection.

Practical Aspects of Transformation in Molecular Biology

The discovery of artificially induced competence in bacteria allow bacteria such as *Escherichia coli* to be used as a convenient host for the manipulation of DNA as well as expressing proteins. Typically plasmids are used for transformation in *E. coli*. In order to be stably maintained in the cell, a plasmid DNA molecule must contain an origin of replication, which allows it to be replicated in the cell independently of the replication of the cell's own chromosome.

The efficiency with which a competent culture can take up exogenous DNA and express its genes is known as transformation efficiency and is measured in colony forming unit (cfu) per µg DNA used. A transformation efficiency of 1×10^8 cfu/µg for a small plasmid like pUC19 is roughly equivalent to 1 in 2000 molecules of the plasmid used being transformed.

In calcium chloride transformation, the cells are prepared by chilling cells in the presence of Ca^{2+} (in $CaCl_2$ solution), making the cell become permeable to plasmid DNA. The cells are incubated on ice with the DNA, and then briefly heat-shocked (e.g., at 42 °C for 30–120 seconds). This method works very well for circular plasmid DNA. Non-commercial preparations should normally give 10^6 to 10^7 transformants per microgram of plasmid; a poor preparation will be about 10^4/µg or less, but a good preparation of competent cells can give up to ~10^8 colonies per microgram of plasmid. Protocols, however, exist for making supercompetent cells that may yield a transformation efficiency of over 10^9. The chemical method, however, usually does not work well for linear DNA, such as fragments of chromosomal DNA, probably because the cell's native exonuclease enzymes rapidly degrade linear DNA. In contrast, cells that are naturally competent are usually transformed more efficiently with linear DNA than with plasmid DNA.

The transformation efficiency using the $CaCl_2$ method decreases with plasmid size, and electroporation therefore may be a more effective method for the uptake of large plasmid DNA. Cells used in electroporation should be prepared first by washing in cold double-distilled water to remove charged particles that may create sparks during the electroporation process.

Selection and Screening in Plasmid Transformation

Because transformation usually produces a mixture of relatively few transformed cells and an abundance of non-transformed cells, a method is necessary to select for the cells that have acquired the plasmid. The plasmid therefore requires a selectable marker such that those cells without the plasmid may be killed or have their growth arrested. Antibiotic resistance is the most commonly used marker for prokaryotes. The transforming plasmid contains a gene that confers resistance to an antibiotic that the bacteria are otherwise sensitive to. The mixture of treated cells is cultured on media that contain the antibiotic so that only transformed cells are able to grow. Another method of selection is the use of certain auxotrophic markers that can compensate for an inability to metabolise certain amino acids, nucleotides, or sugars. This method requires the use of suitably mutated strains that are deficient in the synthesis or utility of a particular biomolecule, and the transformed cells are cultured in a medium that allows only cells containing the plasmid to grow.

In a cloning experiment, a gene may be inserted into a plasmid used for transformation. However, in such experiment, not all the plasmids may contain a successfully inserted gene. Additional techniques may therefore be employed further to screen for transformed cells that contain plasmid with the insert. Reporter genes can be used as markers, such as the *lacZ* gene which codes for β-galactosidase used in blue-white screening. This method of screening relies on the principle of α-complementation, where a fragment of the *lacZ* gene (*lacZα*) in the plasmid can complement another mutant *lacZ* gene (*lacZΔM15*) in the cell. Both genes by themselves produce non-functional peptides, however, when expressed together, as when a plasmid containing *lacZ-α* is transformed into a *lacZΔM15* cells, they form a functional β-galactosidase. The presence of an active β-galactosidase may be detected when cells are grown in plates containing X-gal, forming characteristic blue colonies. However, the multiple cloning site, where a gene of interest may be ligated into the plasmid vector, is located within the *lacZα* gene. Successful ligation therefore disrupts the *lacZα* gene, and no functional β-galactosidase can form, resulting in white colonies. Cells containing successfully ligated insert can then be easily identified by its white coloration from the unsuccessful blue ones.

Other commonly used reporter genes are green fluorescent protein (GFP), which produces cells that glow green under blue light, and the enzyme luciferase, which catalyzes a reaction with luciferin to emit light. The recombinant DNA may also be detected using other methods such as nucleic acid hybridization with radioactive RNA probe, while cells that expressed the desired protein from the plasmid may also be detected using immunological methods.

Natural Competence

In microbiology, genetics, cell biology, and molecular biology, competence is the ability of a cell to take up extracellular ("naked") DNA from its environment in the process called transformation. Competence may be differentiated between *natural competence*, a genetically specified ability of bacteria which is thought to occur under natural conditions as well as in the laboratory, and *induced* or *artificial competence*, which arises when cells in laboratory cultures are treated to make them transiently permeable to DNA. Here we primarily deal with natural competence in bacteria, although information about artificial competence is also provided.

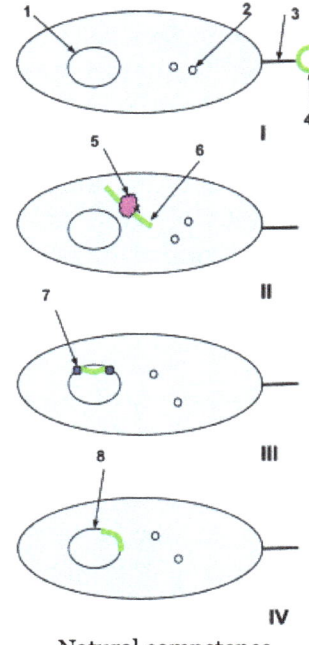

1-Bacterial cell DNA

2-Bacterial cell plasmids

3-Sex pili

4-Plasmid of foreign DNA from a dead cell

5-Bacterial cell restriction enzyme

6-Unwound foreign plasmid7-DNA ligase

I: A plasmid of foreign DNA from a dead cell is intercepted by the sex pili of a naturally competent bacterial cell.

II: The foreign plasmid is transduced through the sex pili into the bacterial cell, where it is processed by bacterial cell restriction enzymes. The restriction enzymes break the foreign plasmid into a strand of nucleotides that can be added to the bacterial DNA.

III: DNA ligase integrates the foreign nucleotides into the bacterial cell DNA.

IV: Recombination is complete and the foreign DNA has integrated into the original bacterial cell's DNA and will continue to be a part of it when the bacterial cell replicates next.

Natural competence.

History

Natural competence was discovered by Frederick Griffith in 1928, when he showed that a preparation of killed cells of a pathogenic bacterium contained something that could transform related non-pathogenic cells into the pathogenic type. In 1944 Oswald Avery, Colin MacLeod, and Maclyn McCarty demonstrated that this 'transforming factor' was pure DNA . This was the first compelling evidence that DNA carries the genetic information of the cell.

Since then, natural competence has been studied in a number of different bacteria, particularly *Bacillus subtilis*, *Streptococcus pneumoniae* (Griffith's "pneumococcus"), *Neisseria gonorrhoeae* and *Haemophilus influenzae*. Areas of active research include the mechanisms of DNA transport, the regulation of competence in different bacteria, and the evolutionary function of competence.

Mechanisms of DNA Uptake

In the natural world DNA usually becomes available by death and lysis of other cells, but in the laboratory it is provided by the researcher, often as a genetically engineered fragment or plasmid. During uptake, DNA is transported across the cell membrane(s), and the cell wall if one is present. Once the DNA is inside the cell it may be degraded to nucleotides, which are reused for DNA replication and other metabolic functions. Alternatively it may be recombined into the cell's genome by its DNA repair enzymes. If this recombination changes the cell's genotype the cell is said to have been transformed. Artificial competence and transformation are used as research tools in many organisms *(genetics)*.

In almost all naturally competent bacteria components of extracellular filaments called type 4 pili (a type of fimbria) bind extracellular double stranded DNA. The DNA is then translocated across the membrane (or membranes for gram negative bacteria) through multi-component protein complexes driven by the degradation of one strand of the DNA. Single stranded DNA in the cell is bound by a well-conserved protein, DprA, which loads the DNA onto RecA, which mediates homologous recombination through the classic DNA repair pathway.

Regulation of Competence

In laboratory cultures, natural competence is usually tightly regulated and often triggered by nutritional shortages or adverse conditions. However the specific inducing signals and regulatory machinery are much more variable than the uptake machinery, and little is known about the regulation of competence in the natural environments of these bacteria. Transcription factors have been discovered which regulate competence; an example is sxy (also known as tfoX) which has been found to be regulated in turn by a 5' non-coding RNA element. In bacteria capable of forming spores, conditions inducing sporulation often overlap with those inducing competence. Thus cultures or colonies containing sporulating cells often also contain competent cells. Recent research by Süel *et al*. has identified an excitable core module of genes which can explain entry into and exit from competence when cellular noise is taken into account.

Most competent bacteria are thought to take up all DNA molecules with roughly equal efficiencies, but bacteria in the families Neisseriaceae and Pasteurellaceae preferentially take up DNA fragments containing short DNA sequences, termed DNA uptake sequence (DUS and USS respectively), that are very frequent in their own genomes. Neisserial genomes contain thousands of copies of the preferred sequence GCCGTCTGAA, and Pasteurellacean genomes contain either AAGTGCGGT or ACAAGCGGT.

Evolutionary Functions and Consequences of Competence

Most proposals made for the primary evolutionary function of natural competence as a part of natural bacterial transformation fall into three categories: (1) the selective advantage of genetic diversity; (2) DNA uptake as a source of nucleotides (DNA as "food"); and (3) the selective advantage of a new strand of DNA to promote homologous recombinational repair of damaged DNA (DNA repair). A secondary suggestion has also been made, noting the occasional advantage of horizontal gene transfer.

Hypothesis of Genetic Diversity

Arguments to support genetic diversity as the primary evolutionary function of sex (including bacterial transformation) are given by Barton and Charleworth . and by Otto and Gerstein. However, the theoretical difficulties associated with the evolution of sex suggest that sex for genetic diversity is problematic. Specifically with respect to bacterial transformation, competence requires the high cost of a global protein synthesis switch, with, for example, more than 16 genes that are switched on only during competence of *Streptococcus pneumoniae*. However, since bacteria tend to grow in clones, the DNA available for transformation would generally have the same genotype as that of the recipient cells. Thus, there is always a high cost in protein expression without, in general, an increase in diversity. Other differences between competence and sex have been considered in models of the evolution of genes causing competence; these models found that competence's postulated recombinational benefits were even more elusive than those of sex.

Hypothesis of DNA as Food

The second hypothesis, DNA as food, relies on the fact that cells that take up DNA inevitably acquire the nucleotides the DNA consists of, and, because nucleotides are needed for DNA and RNA synthesis and are expensive to synthesize, these may make a significant contribution to the cell's energy budget. Some naturally competent bacteria also secrete nucleases into their surroundings, and all bacteria can take up the free nucleotides these nucleases generate from environmental DNA. The energetics of DNA uptake are not understood in any system, so it is difficult to compare the efficiency of nuclease secretion to that of DNA uptake and internal degradation. In principle the cost of nuclease production and the uncertainty of nucleotide recovery must be balanced against the energy needed to synthesize the uptake machinery and to pull DNA in. Other important factors are the likelihoods that nucleases and competent cells will encounter DNA molecules, the relative inefficiencies of nucleotide uptake from the environment and from the periplasm (where one strand is degraded by competent cells), and the advantage of producing ready-to-use nucleotide monophosphates from the other strand in the cytoplasm. Another complicating factor is the self-bias of the DNA uptake systems of species in the family *Pasteurellaceae* and the genus *Neisseria*, which could reflect either selection for recombination or for mechanistically efficient uptake.

Hypothesis of Repair of DNA Damage

In bacteria, the problem of DNA damage is most pronounced during periods of stress, particularly oxidative stress, that occur during crowding or starvation conditions. Under such conditions there is often only a single chromosome present. The finding that some bacteria induce competence under such stress conditions, supports the third hypothesis, that transformation exists to permit DNA repair. In experimental tests, bacterial cells exposed to agents damaging their DNA, and then undergoing transformation, survived better than cells exposed to DNA damage that did not undergo transformation (Hoelzer and Michod, 1991). In addition, competence to undergo transformation is often inducible by known DNA damaging agents (reviewed by Michod *et al.*, 2008 and Bernstein *et al.*, 2012). Thus, a strong short-term selective advantage for natural competence and transformation would be its ability to promote homologous recombinational DNA repair under conditions of stress. Such stress conditions might be incurred during bacterial infection of a susceptible host. Consistent with this idea, Li et al. reported that, among different highly transformable *S. pneumoniae* isolates, nasal colonization fitness and virulence (lung infectivity) depends on an intact competence system.

A counter argument was made based on the 1993 report of Redfield who found that single-stranded and double-stranded damage to chromosomal DNA did not induce or enhance competence or transformation in *B. subtilis* or *H. influenzae*, suggesting that selection for repair has played little or no role in the evolution of competence in these species.

However more recent evidence indicates that competence for transformation is, indeed, specifically induced by DNA damaging conditions. For instance, Claverys *et al.* in 2006 showed that the DNA damaging agents mitomycin C (a DNA cross-linking agent) and fluoroquinolone (a topoisomerase inhibitor that causes double-strand breaks) induce transformation in *Streptococcus pneumoniae*. In addition, Engelmoer and Rozen in 2011 demonstrated that in *S. pneumoniae* transformation protects against the bactericidal effect of mitomycin C. Induction of competence further protected against the antibiotics kanomycin and streptomycin. Although these aminoglycoside antibiotics were previously regarded as non-DNA damaging, recent studies in 2012 of Foti *et al.* showed that a substantial portion of their bactericidal activity results from release of the hydroxyl radical and induction of DNA damages, including double-strand breaks.

Dorer *et al.*, in 2010, showed that ciprofloxacin, which interacts with DNA gyrase and causes production of double-strand breaks, induces expression of competence genes in *Helicobacter pylori*, leading to increased transformation. In 2011 studies of *Legionella pneumophila*, Charpentier *et al.* tested 64 toxic molecules to determine which ones induce competence. Only six of these molecules, all DNA damaging agents, strongly induced competence. These molecules were norfloxacin, ofloxacin and nalidixic acid (inhibitors of DNA gyrase that produce double strand breaks), mitomycin C (which produces inter-strand cross-links), bicyclomycin (causes single- and double-strand breaks), and hydroxyurea (causes oxidation of DNA bases). Charpentier *et al.* also showed that UV irradiation induces competence in *L. pneumophila* and further suggested that competence for transformation evolved as a response to DNA damage.

Horizontal Gene Transfer

A long-term advantage may occasionally be conferred by occasional instances of horizontal gene transfer also called *lateral gene transfer*, (which might result from non-homologous recombination after competence is induced), that could provide for antibiotic resistance or other advantages.

Regardless of the nature of selection for competence, the composite nature of bacterial genomes provides abundant evidence that the horizontal gene transfer caused by competence contributes to the genetic diversity that makes evolution possible.

Agrobacterium

Agrobacterium is a genus of Gram-negative bacteria established by H. J. Conn that uses horizontal gene transfer to cause tumors in plants. *Agrobacterium tumefaciens* is the most commonly studied species in this genus. *Agrobacterium* is well known for its ability to transfer DNA between itself and plants, and for this reason it has become an important tool for genetic engineering.

The *Agrobacterium* genus is quite heterogeneous. Recent taxonomic studies have reclassified all of the *Agrobacterium* species into new genera, such as *Ahrensia*, *Pseudorhodobacter*, *Ruegeria*, and *Stappia*, but most species have been controversially reclassified as *Rhizobium* species.

Plant Pathogen

A. tumefaciens causes crown-gall disease in plants. The disease is characterised by a tumour-like growth or gall on the infected plant, often at the junction between the root and the shoot. Tumors are incited by the conjugative transfer of a DNA segment (T-DNA) from the bacterial tumour-inducing (Ti) plasmid. The closely related species, *A. rhizogenes*, induces root tumors, and carries the distinct Ri (root-inducing) plasmid. Although the taxonomy of *Agrobacterium* is currently under revision it can be generalised that 3 biovars exist within the genus, *A. tumefaciens*, *A. rhizogenes*, and *A. vitis*. Strains within *A. tumefaciens* and *A. rhizogenes* are known to be able to harbour either a Ti or Ri-plasmid, whilst strains of *A. vitis*, generally restricted to grapevines, can harbour a Ti-plasmid. Non-*Agrobacterium* strains have been isolated from environmental samples which harbour a Ri-plasmid whilst laboratory studies have shown that non-*Agrobacterium* strains can also harbour a Ti-plasmid. Some environmental strains of *Agrobacterium* possess neither a Ti nor Ri-plasmid. These strains are avirulent.

The large growths on these roots are galls induced by *Agrobacterium* sp.

The plasmid T-DNA is integrated semi-randomly into the genome of the host cell, and the tumor morphology genes on the T-DNA are expressed, causing the formation of a gall. The T-DNA carries genes for the biosynthetic enzymes for the production of unusual amino acids, typically octopine or nopaline. It also carries genes for the biosynthesis of the plant hormones, auxin and cytokinins, and for the biosynthesis of opines, providing a carbon and nitrogen source for the bacteria that most other micro-organisms can't use, giving *Agrobacterium* a selective advantage. By altering the hormone balance in the plant cell, the division of those cells cannot be controlled by the plant, and tumors form. The ratio of auxin to cytokinin produced by the tumor genes determines the morphology of the tumor (root-like, disorganized or shoot-like).

In Humans

Although generally seen as an infection in plants, *Agrobacterium* can be responsible for opportunistic infections in humans with weakened immune systems, but has not been shown to be a primary pathogen in otherwise healthy individuals. One of the earliest associations of human disease caused by *Agrobacterium radiobacter* was reported by Dr. J. R. Cain in Scotland (1988). A later study suggested that *Agrobacterium* attaches to and genetically transforms several types of human cells by integrating its T-DNA into the human cell genome. The study was conducted using cultured human tissue and did not draw any conclusions regarding related biological activity in nature.

Uses in Biotechnology

The ability of *Agrobacterium* to transfer genes to plants and fungi is used in biotechnology, in particular, genetic engineering for plant improvement. A modified Ti or Ri plasmid can be used.

The plasmid is 'disarmed' by deletion of the tumor inducing genes; the only essential parts of the T-DNA are its two small (25 base pair) border repeats, at least one of which is needed for plant transformation. The genes to be introduced into the plant are cloned into a plant transformation vector that contains the T-DNA region of the disarmed plasmid, together with a selectable marker (such as antibiotic resistance) to enable selection for plants that have been successfully transformed. Plants are grown on media containing antibiotic following transformation, and those that do not have the T-DNA integrated into their genome will die. An alternative method is agroinfiltration.

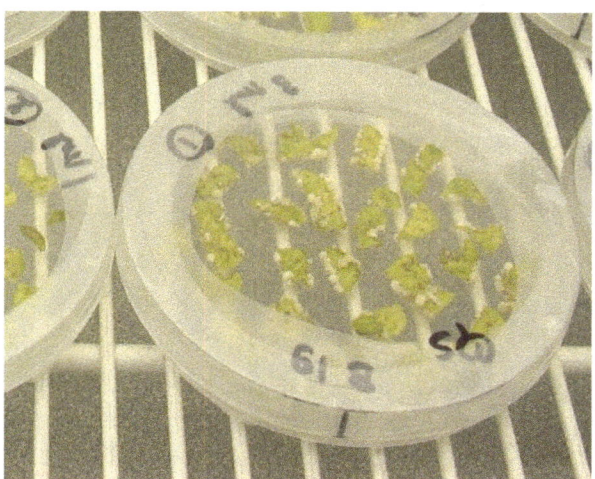

Plant (*S. chacoense*) transformed using *Agrobacterium*. Transformed cells start forming calluses on the side of the leaf pieces

Transformation with *Agrobacterium* can be achieved in two ways. Protoplasts or alternatively leaf-discs can be incubated with the *Agrobacterium* and whole plants regenerated using plant tissue culture. A common transformation protocol for *Arabidopsis* is the floral-dip method: inflorescence are dipped in a suspension of *Agrobacterium*, and the bacterium transforms the germline cells that make the female gametes. The seeds can then be screened for antibiotic resistance (or another marker of interest), and plants that have not integrated the plasmid DNA will die when exposed to the correct condition of antibiotic.

Agrobacterium does not infect all plant species, but there are several other effective techniques for plant transformation including the gene gun.

Agrobacterium is listed as being the vector of genetic material that was transferred to these USA GMOs:

- Soybean
- Cotton
- Corn
- Sugar Beet
- Alfalfa
- Wheat

- Rapeseed Oil (Canola)

- Creeping bentgrass (for animal feed)

- Rice (Golden Rice)

Genomics

The sequencing of the genomes of several species of *Agrobacterium* has permitted the study of the evolutionary history of these organisms and has provided information on the genes and systems involved in pathogenesis, biological control and symbiosis. One important finding is the possibility that chromosomes are evolving from plasmids in many of these bacteria. Another discovery is that the diverse chromosomal structures in this group appear to be capable of supporting both symbiotic and pathogenic lifestyles. The availability of the genome sequences of *Agrobacterium* species will continue to increase, resulting in substantial insights into the function and evolutionary history of this group of plant-associated microbes.

History

Marc Van Montagu and Jozef Schell at the University of Ghent (Belgium) discovered the gene transfer mechanism between *Agrobacterium* and plants, which resulted in the development of methods to alter *Agrobacterium* into an efficient delivery system for gene engineering in plants. A team of researchers led by Dr Mary-Dell Chilton were the first to demonstrate that the virulence genes could be removed without adversely affecting the ability of *Agrobacterium* to insert its own DNA into the plant genome (1983).

Viral Transformation

Viral transformation is the change in growth, phenotype, or indefinite reproduction of cells caused by the introduction of inheritable material. Through this process, a virus causes harmful transformations of an in vivo cell or cell culture. The term can also be understood as DNA transfection using a viral vector.

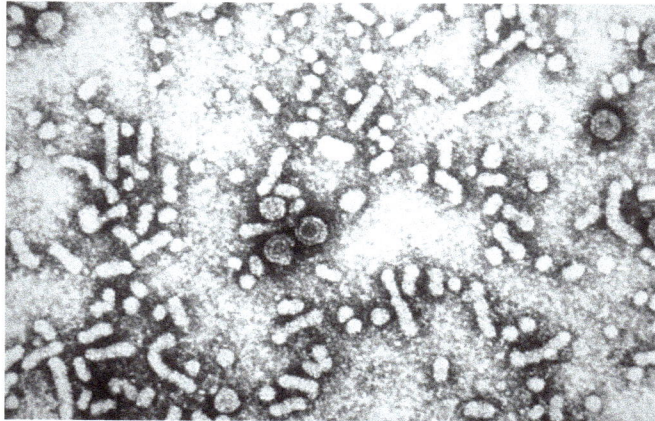

Hepatitis-B virions

Viral transformation can occur both naturally and medically. Natural transformations can include viral cancers, such as human papillomavirus (HPV) and T-cell Leukemia virus type I. Hepatitis B and C are also the result of natural viral transformation of the host cells. Viral transformation can also be induced for use in medical treatments.

Cells that have been virally transformed can be differentiated from untransformed cells through a variety of growth, surface, and intracellular observations. The growth of transformed cells can be impacted by a loss of growth limitation caused by cell contact, less oriented growth, and high saturation density. Transformed cells can lose their tight junctions, increase their rate of nutrient transfer, and increase their protease secretion. Transformation can also affect the cytoskeleton and change in the quantity of signal molecules.

Type

There are three types of viral infections that can be considered under the topic of viral transformation. These are cytocidal, persistent, and transforming infections. Cytocidal infections can cause fusion of adjacent cells, disruption of transport pathways including ions and other cell signals, disruption of DNA, RNA and protein synthesis, and nearly always leads to cell death. Persistent infections involve viral material that lays dormant within a cell until activated by some stimulus. This type of infection usually causes few obvious changes within the cell but can lead to long chronic diseases. Transforming infections are also referred to as malignant transformation. This infection causes a host cell to become malignant and can be either cytocidal (usually in the case of RNA viruses) or persistent (usually in the case of DNA viruses). Cells with transforming infections undergo immortalization and inherit the genetic material to produce tumors. Since the term cytocidal, or cytolytic, refers to cell death, these three infections are not mutually exclusive. Many transforming infections by DNA tumor viruses are also cytocidal.

Table 1: Cellular effects of viral infections

		Genetic	Cell Fate	Morpholog-ical	Biochemical	Physiological
Cytocidal	Productive / ----	DNA degradation / ----	Death / ----	Rounding of the cell Fusion with adjacent cells Appearance of inclusion bodies	Inhibit DNA, RNA, and protein synthesis Interfere with sub-cellular interactions	Insufficient movement of ions Formation of secondary messengers Activation of cellular cascades
	Abortive	Possible mutation	Usually death			
Persistent	Latent / ----	Possible Mutation / ----	Survival / ----	Fusion with adjacent cells Appearance of inclusion bodies Budding	Immune responses limit viral spread Antigen-antibody complexes can incorporate viral antigens causing inflammation	Rare until stimulated
	Chronic / ----	Possible Mutation / ----	Variable / ----			
	Slow	Possible Mutation	Variable			

Transforming	DNA viruses ------------ ---- RNA viruses	Mutation ------------ ---- Mutation	Survival ------------ ---- Survival	Unlimited cell replication	Inactivates tumor suppressor proteins Impairs cell cycle regulation	Unlimited cell replication

Cytocidal Infections

Cytocidal infections are often associated with changes in cell morphology, physiology and are thus important for the complete viral replication and transformation. *Cytopathic Effects*, often include a change in cell's morphology such as fusion with adjacent cells to form polykaryocytes as well as the synthesis of nuclear and cytoplasmic inclusion bodies. *Physiological changes* include the insufficient movement of ions, formation of secondary messengers, and activation of cellular cascades to continue cellular activity. *Biochemically*, many viruses inhibit the synthesis of host DNA, RNA, proteins directly or even interfere with protein-protein, DNA-protein, RNA-protein interactions at the subcellular level. *Genotoxicity* involves breaking, fragmenting, or rearranging chromosomes of the host. Lastly, *biologic effects* include the viruses' ability to affect the activity of antigens and immunologlobulins in the host cell.

There are two types of cytocidal infections, productive and abortive. In productive infections, additional infectious viruses are produced. Abortive infections do not produce infectious viruses. One example of a productive cytocidal infection is the herpes virus.

Persistent Infections

There are three types of persistent infections, latent, chronic and slow, in which the virus stays inside the host cell for prolonged periods of time. During *latent infections* there is minimal to no expression of infected viral genome. The genome remains within the host cell until the virus is ready for replication. *Chronic infections* have similar cellular effects as acute cytocidal infections but there is a limited number of progeny and viruses involved in transformation. Lastly, *slow infections* have a longer incubation period in which no physiological, morphological or subcellular changes may be involved.

Transforming Infections

Transformation infections is limited to abortive or restrictive infections. This constitutes the broadest category of infections as it can include both cytocidal and persistent infection. Viral transformation is most commonly understood as transforming infections.

Process

In order for a cell to be transformed by a virus, the viral DNA must be entered into the host cell. The simplest consideration is viral transformation of a bacterial cell. This process is called lysogeny. As shown in the above figure, a bacteriophage lands on a cell and pins itself to the cell. The phage can then penetrate the cell membrane and inject the viral DNA into the host cell. The viral DNA can then either lay dormant until stimulated by a source such as UV light or it can be immediately taken up by the host's genome. In either case the viral DNA will replicate along with the original host DNA during cell replication causing two cells to now be infected with the virus. The process will continue to propagate more and more infected cells. This process is in contrast to the lytic cycle where a virus only uses the host cell's replication machinery to replicate itself before destroying the host cell.

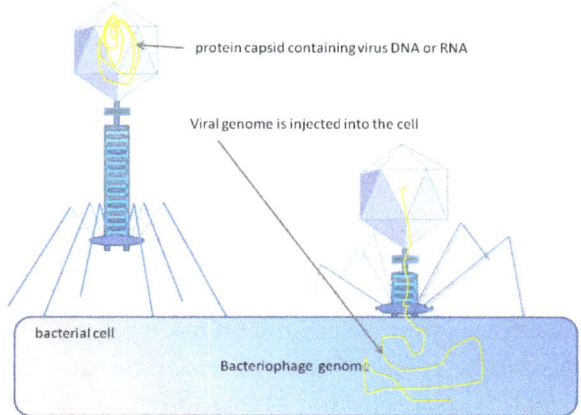

Phage injecting its genome into bacterial cell

The process is similar in animal cells. In most cases, rather than viral DNA being injected into an animal cell, a section of the membrane encases the virus and the cell then absorbs both the virus and the encasing section of the membrane into the cell. This process, called endocytosis, is shown in the figure below.

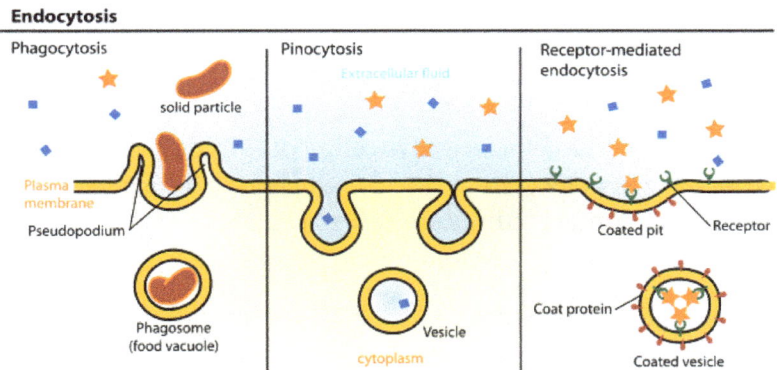

Examples of endocytosis

Transformation of the Host Cell

Viral transformation disrupts the normal expression of the host cell's genes in favor of expressing a limited number of viral genes. The virus also can disrupt communication between cells and cause cells to divide at an increased rate.

Physiological

Viral transformation can impose characteristically determinable features upon a cell. Typical phenotypic changes include high saturation density, anchorage-independent growth, loss of contact inhibition, loss of orientated growth, immortalization, disruption of the cell's cytoskeleton.

Biochemical

Viral genes are expressed through the use of the host cell's replication machinery; therefore, many viral genes have promoters that support binding of many transcription factors found naturally in the host cells. These transcription factors along with the virus' own proteins can repress or activate genes from both the virus and the host cell's genome. Many viruses can also increase the production of the cell's regulatory proteins.

Genetic

Depending on the virus, a variety of genetic changes can occur in the host cell. In the case of a lytic cycle virus, the cell will only survive long enough to the replication machinery to be used to create additional viral units. In other cases, the viral DNA will persist within the host cell and replicate as the cell replicates. This viral DNA can either be incorporated into the host cell's genetic material or persist as a separate genetic vector. Either case can lead to damage of the host cell's chromosomes. It is possible that the damage can be repaired; however, the most common result is an instability in the original genetic material or suppression or alteration of the gene expression.

Assays

An assay is an analytic tool often used in a laboratory setting in order to assess or measure some quality of a target entity. In virology, assays can be used to differentiate between transformed and non-transformed cells. Varying the assay used, changes the selective pressure on the cells and therefore can change what properties are selected in the transformed cells.

Three common assays used are the focus forming assay, the Anchorage independent growth assay, and the reduced serum assay.

The focus forming assay (FFA) is used to grow cells containing a transforming oncogene on a monolayer of non-transformed cells. The transformed cells will form raised, dense spots on the sample as they grow without contact inhibition. This assay is highly sensitive compared to other assays used for viral analysis, such as the yield reduction assay.

An example of the Anchorage independent growth assay is the soft agar assay. The assay is assessing the cells' ability to grow in a gel or viscous fluid. Transformed cells can grow in this environment and are considered anchorage independent. Cells that can only grow when attached to a solid surface are anchorage dependent untransformed cells. This assay is considered one of the most stringent for detection of malignant transformation

In a reduced serum assay, cells are assayed by exploiting the changes in cell serum requirements. Non-transformed cells require at least a 5% serum medium in order to grow; however, transformed cells can grow in an environment with significantly less serum.

Examples of Natural Transformation

Natural transformation is the viral transformation of cells without the interference of medical science. This is the most commonly considered form of viral transformation and includes many cancers and diseases, such as HIV, Hepatitis B, and T-cell Leukemia virus type I.

Viral Oncogenesis

As many as 20% of human tumors are caused by viruses. Some such viruses that are commonly recognized include HPV, T-cell Leukemia virus type I, and hepatitis B.

Viral oncogenesis are most common with DNA and RNA tumor viruses, most frequently the retroviruses. There are two types of oncogenic retroviruses: acute transforming viruses and non-acute transforming viruses. Acute transforming viruses induce a rapid tumor growth since they carry viral oncogenes in their DNA/RNA to induce such growth. An example of an acute transforming virus is the Rous Sarcoma Virus (RSV) that carry the v-src oncogene. v-Src is part of the c-src, which is a cellular proto-oncogene that stimulates rapid cell growth and expansion. A non-acute transforming virus on the other hand induces a slow tumor growth, since it does not carry any viral oncogenes. It induces tumor growth by transcriptionally activating the proto-oncogenes particularly the long terminal repeat (LTR) in the proto-oncogenes.

Viral Oncogonesis through transformation can occur via 2 mechanisms:

1. The tumor virus can introduce and express a "transforming" gene either through the integration of DNA or RNA into the host genome.

2. The tumor virus can alter expression on preexisting genes of the host.

One or both of these mechanisms can occur in the same host cell.

Hepatitis B

The Hepatitis B viral protein X is believed to cause hepatocellular carcinoma through transformation, typically of liver cells. The viral DNA is incorporated into the host cell's genome causing rapid cell replication and tumor growth.

Papillomaviruses

Papillomaviruses typically target epithelial cells and cause everything from warts to cervical cancer. When human papillomavirus (HPV) transforms a cell, it interferes with the function of cellular proteins while degrading other cellular proteins.

Herpesviruses

The herpesviruses, Kaposi's sarcoma-associated herpesvirus and Epstein-Barr virus, are believed to cause cancer in humans, such as Kaposi's sarcoma, Burkitt's lymphoma, and nasopharyngeal carcinoma. Although genes have been identified in these viruses that cause transformation, the manner in which the virus transforms and replicates the host cell is not understood.

Retroviruses

The retroviruses include T-cell Leukemia virus type I, HIV, and Rous Sarcoma Virus (RSV). The viral gene tax is expressed when the T-cell Leukemia virus transforms a cell altering the expression of cellular growth control genes and causing the transformed cells to become cancerous. HIV works differently by not directly causing cells to become cancerous but by instead making those infected more susceptible to lymphoma and Kaposi's sarcoma. Many other retroviruses contain the three genes, gag, pol, and env, which do not directly cause transformation or tumor formation.

HIV

Human immunodeficiency virus is a viral infection that targets the lymph nodes. HIV binds to the immune CD4 cell and reverse transcriptase alters the host cell genome to allow integration of the viral DNA via integrase. The virus replicates using the host cell's machinery and then leaves the cell to infect additional cells via budding.

Medical Applications

There are many applications in which viral transformation can be artificially induced in a cell culture in order to treat an illness or other condition. A cell culture is infected with a virus causing the transformation; transformed cells can then be used to either produce treatments or be directly introduced into the body.

Personalized Type I Interferons

Type I interferons (IFNs) are used to treat a wide variety of medical conditions including hepatitis C, cancers, viral and inflammatory diseases. IFNs can either be extracted from a natural source, such as cultured human cells or blood leukocytes, or they can be manufactured with recombinant DNA technologies. Most of these IFN treatments have a low response rate.

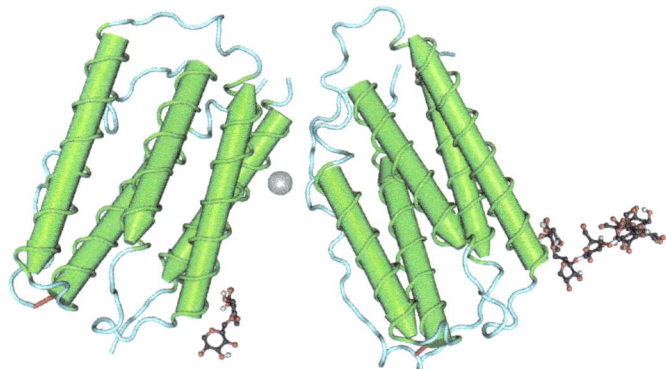

Type I Human Interferon

The use of viral transformation of the Epstein-Barr virus (EBV) has been recommended to create personalized IFNs. In this process, primary B lymphocytes are transformed with EBV. These cells can then be used to produce IFNs specific for the patient from which the B lymphocytes were extracted. This personalization decreases the likelihood of an antibody response and therefore increases the effectiveness of the treatment.

Cancer Treatments

When a virus transforms a cell it often causes cancer by either altering the cells' existing genome or introducing additional genetic material which causes cells to uncontrollably replicate. It is rarely considered that what causes so much harm also has the capability of reversing the process and slowing the cancer growth or even leading to remission. Viruses transform host cells in order to survive and replicate; however, the immune responses of the host cell are typically compromised during transformation making transformed cells more susceptible to other viruses.

The idea of using viruses to treat cancers was first introduced in 1951 when a 4-year-old boy suddenly went into a temporary remission from leukemia while he had chickenpox. This led to research in the 1990s where scientists worked to create a strain of the herpes simplex virus strong enough to infect and transform tumor cells but weak enough to leave healthy cells unharmed. Treating patients with viral transformation has the possibility of treating patients more safely and more effectively than using traditional methods, such as chemotherapy. Viruses used in the treatment of cancer gain strength and increase their effectiveness as the multiply in the body while causing only minor side effects, such as nausea, fatigue, and aches.

Transformation Efficiency

Transformation efficiency is the efficiency by which cells can take up extracellular DNA and express genes encoded by it. This is based on the competence of the cells. It can be calculated by dividing the number of successful transformants by the amount of DNA used during a transformation procedure. Transformants are cells that have taken up DNA (foreign, artificial or modified) and which can express genes on the introduced DNA.

Measurement

Transformation efficiency should be determined under conditions of cell excess. The number of viable cells in a preparation for a transformation reaction may range from 2×10^8 to 10^{11}; most common methods of E. coli preparation yield around 10^{10} viable cells per reaction. The standard plasmids used for determination of transformation efficiency in *Escherichia coli* are pBR322 or other similarly-sized or smaller vectors, such as the pUC series of vectors. Different vectors however may be used to determine their transformation efficiency. 10–100 pg of DNA may be used for transformation, more DNA may be necessary for low-efficiency transformation (generally saturation level is reached at over 10 ng).

After transformation, 1% and 10% of the cells are plated separately, the cells may be diluted in media as necessary for ease of plating. Further dilution may be used for high efficiency transformation.

Transformation efficiency can be measured in transformants or colony forming unit (cfu) per μg DNA used. A transformation efficiency of 1×10^8 cfu/μg for a small plasmid like pUC19 is roughly equivalent to 1 in 2000 molecules of the plasmid used being transformed. In *E. coli*, the theoretical limit of transformation efficiency for most commonly used plasmids would be over 1×10^{11} cfu/μg. In practice the best achievable result may be around 2–4×10^{10} cfu/μg for a small plasmid like pUC19, and considerably lower for large plasmids.

Factors Affecting Transformation Efficiency

Individual cells are capable of taking up many DNA molecules, but the presence of multiple plasmids does not significantly affect the occurrence of successful transformation events. A number of factors may affect the transformation efficiency:

Plasmid size — A study done in *E. coli* found that transformation efficiency declines linearly with increasing plasmid size, i.e. larger plasmids transform less well than smaller plasmids.

Forms of DNA — Supercoiled plasmid have a slightly better transformation efficiency than relaxed plasmids - relaxed plasmids are transformed at around 75% efficiency of supercoiled ones. Linear and single-stranded DNA however have much lower transformation efficiency. Single-stranded DNAs are transformed at 10^4 lower efficiency than double-stranded ones.

Genotype of cells — Cloning strains may contain mutations that improve the transformation efficiency of the cells. For example, *E. coli* K12 strains with the *deoR* mutation, originally found to confer an ability of cell to grow in minimum media using inosine as the sole carbon source, have 4-5 times the transformation efficiency of similar strains without. For linear DNA, which is poorly transformed in *E. coli*, the *recBC* or *recD* mutation can significantly improve the efficiency of its transformation.

Growth of cells — *E. coli* cells are more susceptible to be made competent when it is growing rapidly, cells are therefore normally harvested in the early log phase of cell growth when preparing competent cells. The optimal optical density for harvesting cells normally lies around 0.4, although it may vary with different cell strains. A higher value of 0.94-0.95 has also been found to produce good yield of competent cells, but this can be impractical when cell growth is rapid.

Methods of transformation — The method of preparation of competent cells, the length of time of heat shock, temperature of heat shock, incubation time after heat shock, growth medium used, and various additives, all can affect the transformation efficiency of the cells. The presence of contaminants as well as ligase in a ligation mixture can reduce the transformation efficiency in electroporation, and inactivation of ligase or chloroform extraction of DNA may be necessary for electroporation, alternatively only use a tenth of the ligation mixture to reduce the amount of contaminants. Normal preparation of compentent cells can yield transformation efficiency ranging from 10^6 to 10^8 cfu/μg DNA. Protocols for chemical method however exist for making supercompetent cells that may yield a transformation efficiency of over 1×10^9. Electroporation method in general has better transformation efficiency than chemical methods with over 1×10^{10} cfu/μg DNA possible, and it allows large plasmids of 200 kb in size to be transformed.

Damage to DNA - Exposure of DNA to UV radiation in standard preparative agarose gel electrophoresis procedure for as little as 45 seconds can damage the DNA, and this can significantly reduce the transformation efficiency. Adding cytidine or guanosine to the electrophoresis buffer at 1 mM concentration however may protect the DNA from damage. A higher-wavelength UV radiation (365 nm) which cause less damage to DNA should be used if it is necessary work for work on the DNA on a UV transilluminator for an extended period of time. This longer wavelength UV produces weaker fluorescence with the ethidium bromide intercalated into the DNA, therefore if it is necessary to capture images of the DNA bands, a shorter wavelength (302 or 312 nm) UV radiations may be used. Such exposure however should be limited to a very short time if the DNA is to be recovered later for ligation and transformation.

References

- Mandel M, Higa A (October 1970). "Calcium-dependent bacteriophage DNA infection". Journal of Molecular Biology. 53 (1): 159–62. PMID 4922220. doi:10.1016/0022-2836(70)90051-3

- Alberts B, Johnson A, Lewis J, Raff M, Roberts K, Walter P (2002). Molecular Biology of the Cell. New York: Garland Science. p. G:35. ISBN 978-0-8153-4072-0

- Thomson JA. "Genetic Engineering of Plants" (PDF). Biotechnology. Encyclopedia of Life Support Systems. 3. Retrieved 17 July 2016

- Michod RE, Bernstein H, Nedelcu AM (May 2008). "Adaptive value of sex in microbial pathogens" (PDF). Infection, Genetics and Evolution. 8 (3): 267–85. PMID 18295550. doi:10.1016/j.meegid.2008.01.002

- Sanford JC, Klein TM, Wolf ED, Allen N (1987). "Delivery of substances into cells and tissues using a particle bombardment process". Journal of Particulate Science and Technology. 5: 27–37. doi:10.1080/02726358708904533

- Srivastava S (2013). Genetics of Bacteria (PDF). India: Springer-Verlag. ISBN 978-81-322-1089-4. doi:10.1007/978-81-322-1090-0

- Winship, Timothy R (11 Dec 1979). "A Sensitive Method for Quantification of Vesicular Stomatitis Virus Defective Interfering Particles: Focus Forming Assay" (PDF). Journal of General Virology. 48 (1): 237–240. doi:10.1099/0022-1317-48-1-237. Retrieved 25 March 2014

- Inoue H, Nojima H, Okayama H (November 1990). "High efficiency transformation of Escherichia coli with plasmids". Gene. 96 (1): 23–8. PMID 2265755. doi:10.1016/0378-1119(90)90336-P

- Goodgal SH, Herriott RM (July 1961). "Studies on transformations of Hemophilus influenzae. I. Competence". The Journal of General Physiology. 44 (6): 1201–27. PMC 2195138. PMID 13707010. doi:10.1085/jgp.44.6.1201

- Cooper, Geoffrey M. (2000). The cell : a molecular approach (2nd ed.). Washington: ASM Press. ISBN 0878931023

- Provost, Joseph. "Soft Agar Assay for Cology Formation" (PDF). Wallert and Provost Lab. Retrieved 25 March 2014

- Engelmoer, D J; Rozen, D E (2011). "Competence increases survival during stress in Streptococcus pneumoniae". Evolution. 65 (12): 3475–3485. PMID 22133219. doi:10.1111/j.1558-5646.2011.01402.x

- Ito H, Fukuda Y, Murata K, Kimura A (January 1983). "Transformation of intact yeast cells treated with alkali cations". Journal of Bacteriology. 153 (1): 163–8. PMC 217353. PMID 6336730

- Gietz RD, Woods RA (2002). "Transformation of yeast by lithium acetate/single-stranded carrier DNA/polyethylene glycol method". Methods in Enzymology. Methods in Enzymology. 350: 87–96. ISBN 9780121822538. PMID 12073338. doi:10.1016/S0076-6879(02)50957-5

- Xu, Dongsheng; Zhang, Luwen (June 2010). "Viral transformation for production of personalized type I interferons". Biotechnology Journal. 5 (6): 578–581. doi:10.1002/biot.201000038

- Cain, John Raymond (1988). "A case of septicaemia caused by Agrobacterium radiobacter". Journal of Infection. 16 (2): 205–6. PMID 3351321. doi:10.1016/s0163-4453(88)94272-7

Permissions

We would like to thank the editorial team for lending their expertise to make the book truly unique. They have played a crucial role in the development of this book. Without their invaluable contributions this book wouldn't have been possible. They have made vital efforts to compile up to date information on the varied aspects of this subject to make this book a valuable addition to the collection of many professionals and students.

This book was conceptualized with the vision of imparting up-to-date and integrated information in this field. To ensure the same, a matchless editorial board was set up. Every individual on the board went through rigorous rounds of assessment to prove their worth. After which they invested a large part of their time researching and compiling the most relevant data for our readers.

The editorial board has been involved in producing this book since its inception. They have spent rigorous hours researching and exploring the diverse topics which have resulted in the successful publishing of this book. They have passed on their knowledge of decades through this book. To expedite this challenging task, the publisher supported the team at every step. A small team of assistant editors was also appointed to further simplify the editing procedure and attain best results for the readers.

Apart from the editorial board, the designing team has also invested a significant amount of their time in understanding the subject and creating the most relevant covers. They scrutinized every image to scout for the most suitable representation of the subject and create an appropriate cover for the book.

The publishing team has been an ardent support to the editorial, designing and production team. Their endless efforts to recruit the best for this project, has resulted in the accomplishment of this book. They are a veteran in the field of academics and their pool of knowledge is as vast as their experience in printing. Their expertise and guidance has proved useful at every step. Their uncompromising quality standards have made this book an exceptional effort. Their encouragement from time to time has been an inspiration for everyone.

The publisher and the editorial board hope that this book will prove to be a valuable piece of knowledge for students, practitioners and scholars across the globe.

Index

www.ingramcontent.com/pod-product-compliance
Lightning Source LLC
Chambersburg PA
CBHW061242190326
41458CB00011B/3559